Snakes in the Eagle's Nest

A History of Ground Attacks on Air Bases

Alan Vick

Project AIR FORCE

Prepared for the United States Air Force

RAND

PREFACE

In view of the findings of recent RAND research that suggest that few opponents will be able to challenge the U.S. Air Force (USAF) in the air, a RAND study for the USAF, "Countering U.S. Aerospace Power," has been investigating means that future adversaries might pursue to counter U.S. airpower.

As part of that study, the historical effort reported here examines ground attacks on air bases in conflicts between 1940 and 1992. Its purpose is to offer a comprehensive review of attacker objectives and tactics, and of the most effective defensive countermeasures. The insights gained can be related to current air-base-defense doctrine and tactics and should be of interest to Air Force Security Police in training, operations, and policy positions. Additionally, it is hoped that the report will serve as a useful reference for Security Police courses on air base ground defense and for officers researching the history of ground attacks on air bases. The report should also be of interest to Air Force officers in operations and plans (OPLANS) positions who have a broader responsibility for ensuring the availability of airpower as a ready instrument in the defense of U.S. interests. Finally, the special forces, defense analysis, and military history communities should find the report of interest.

The study was conducted as part of the Strategy, Doctrine, and Force Structure program of Project AIR FORCE and was sponsored by the Director of Plans, Headquarters, United States Air Force (AF/XOX).

Project AIR FORCE, a division of RAND, is the Air Force federally funded research and development center (FFRDC) for studies and analysis. It provides the Air Force with independent analyses of policy alternatives affecting the development, employment, combat readiness, and support of current and future aerospace forces. Research is carried out in three programs: Strategy, Doctrine, and Force Structure; Force Modernization and Employment; and Resource Management and System Acquisition.

Project AIR FORCE is operated under Contract F49620-91-C-0003 between the Air Force and RAND.

CONTENTS

FIGURES

TABLES

SUMMARY

Recent RAND research on trends in global airpower suggests that few opponents will be able to challenge the U.S. Air Force (USAF) in the air. If that is correct, future adversaries are likely to look for alternative means to counter U.S. airpower. A RAND study for the Air Force, "Countering U.S. Aerospace Power," has been investigating those means. The historical effort reported here was part of that study and sought to better understand past, present, and future ground threats to air bases.

In the course of the research, it became clear that attacks on air bases were much more frequent and successful than is commonly appreciated. For this reason, the history of those attacks is pertinent to future USAF operations.

This report presents a comprehensive overview of ground attacks on air bases from the first known attacks in 1940 to the most recent in 1992. The objectives, tactics, and outcomes of those attacks are analyzed to identify lessons learned and their applications to future conflicts. In particular, this report identifies the attack techniques that proved most difficult to counter and offers some suggestions for improving air base defenses against them.

BACKGROUND

In 1921, Italian Army General Giulio Douhet observed that "it is easier and more effective to destroy the enemy's aerial power by destroying his nests and eggs on the ground than to hunt his flying birds in the air."[1] Douhet's metaphor was directed at fellow airmen, pointing out both the great offensive potential of airpower—a radical notion in 1921—and the exceptional vulnerability of aircraft on the ground. Flying machines, even modern ones, by their very nature are thin-skinned, relatively soft targets. Speed, maneuverability, and stealth enable these unarmored vehicles to survive and be decisive in combat. In contrast, an aircraft parked on a ramp has none of these characteristics and—compared with most other ground targets—is triflingly easy to destroy. The vulnerability of parked aircraft was vividly demonstrated by the Japanese at Hickam Field, Hawaii, and the demonstration was repeated by all combatants many times during World War II. The preemptive Israeli raid on Egyptian airfields in the 1967 war demonstrated that unsheltered aircraft remain a tempting target in modern air warfare also. Since 1967, billions of dollars have been spent by the world's air forces on aircraft shelters, air defenses, and programs to enhance air base survivability.

Douhet's observation, like most great insights, has applicability beyond the confines of its initial setting. If aircraft are vulnerable on the ground, why not attack them with every weapon available? That is just what the world's armies have done at least 645 times[2] in ten conflicts between 1940 and 1992, destroying or damaging over 2,000 aircraft. Ground attack forces have included airborne, airmobile, infantry, and armor elements. Airborne forces have arrived on the objective by parachute, glider, and aircraft landing, often under fire. Armor and infantry have assaulted by land, and amphibious forces have landed by sea. More recently, helicopters have been used to transport the assault force. Finally, special forces, guerrillas, and terrorists have made their contribution.

---

[1]Giulio Douhet, *The Command of the Air*, Washington, D.C.: Office of Air Force History, 1983 (originally published in 1921), pp. 53–54.

[2]This number is based on *deliberate* attacks on airfields, whether they were independent operations or part of a larger offensive. It does not include the many times that ground forces overran airfields on their way to other objectives.

PURPOSE AND APPROACH

Given the numerous occurrences, global distribution, and recentness of ground attacks on air bases, it is surprising that a history of those attacks has not been published.[3] This report is intended to begin filling that void by bringing together in one document descriptions and analyses of air base attacks over the past 50 years. Beginning with a discussion of the four broad objectives of air base attackers, the report then briefly describes examples of air base attacks under each objective. The core of the report focuses on three case-study regions in which many air base attacks occurred: Crete and North Africa during World War II and Southeast Asia during the Vietnam War. The objectives, tactics, and outcomes of both standoff and penetrating attacks[4] are analyzed to identify lessons learned that can be applied in future conflicts. In particular, by seeking answers to the following questions, I identify those techniques that were most effective for attackers and the successful defensive countermeasures:

- How have attacking forces been inserted into the enemy rear area?
- What attacker tactics and weapons have been most effective?
- What defensive countermeasures have worked?
- Were there promising countermeasures that the defense failed to employ?
- What has been the strategic effect of ground attacks on air bases in previous conflicts?

---

[3]The only published historical work on this subject is Roger Fox, *Air Base Defense in the Republic of Vietnam: 1961–1973*, Washington, D.C.: U.S. Air Force Office of History [now Center for History], 1979. Fox's book is an excellent history of air base defense in Vietnam, but it does not address attacks in other conflicts.

[4]*Standoff attacks* use direct- or indirect-fire weapons from beyond the defensive perimeter. Mortars, rockets, recoilless rifles, and small arms have all been used to fire on aircraft, facilities, and personnel from distances up to 11 kilometers. *Penetrating attacks* typically are done covertly by small teams who slip through the defensive perimeter and place bombs with time fuzes (*satchel charges*) on aircraft and materiel. Defensive perimeters have been assaulted outright also; the attackers then use direct-fire weapons (machine guns, tank guns, and small arms) against airfield targets.

It is hoped that these historical insights will be helpful to USAF officers responsible for planning and executing air base defense today and in the future.

CATEGORIZING AIR BASE ATTACKS

Between 1940 and 1992, ground attacks on air bases pursued a variety of objectives. These objectives ranged from the very ambitious goal of capturing an airfield to the minimalist goal of harassing air base operations. Discussions of air base defense often treat these bounding goals as similar, but they really are quite different and call for a broad range of defensive countermeasures. To make the range and nature of historical threats to air bases more visible, I categorized the attacks identified in this research according to which one of the following four broad categories the attacker's major objective fit best (number of attack type follows each objective):

- Capture airfield (41)
- Deny defender use of airfield (47)
- Harass defenders (173)
- Destroy aircraft and equipment (384).

As Figure S.1 illustrates, the majority (60 percent) of these attacks sought to destroy aircraft and equipment. Only 6 percent were directed at the more ambitious objective of actually capturing airbases as airheads for troop insertion or for offensive use by the attacker's air force. Most of these major attacks occurred during World War II, although Soviet forces in Afghanistan (1979) and U.S. forces in Grenada (1983) and Panama (1989) also seized airfields for use as airheads. With these three exceptions, the most likely threat facing current and future air base defenders appears to be attempts to destroy aircraft rather than to seize the airfield. Figure S.2 shows the distribution of attack tactics for the 645 attacks identified in this report.

Of particular interest is the apparent evolution of air base attacker tactics since World War II (WW II). All the British attacks on Axis airfields in WW II penetrated the defenses. In contrast, faced with extensive minefields, fencing, guard posts, and lights, Viet Cong and

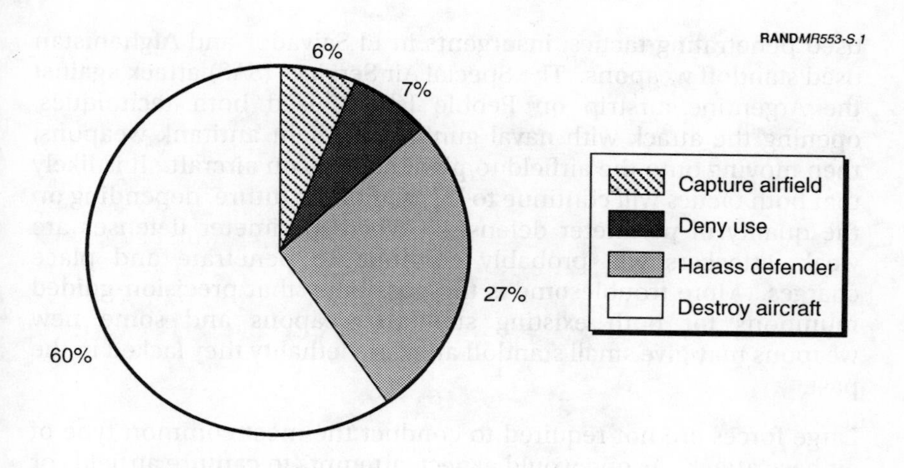

Figure S.1—Airfield-Attack Objectives

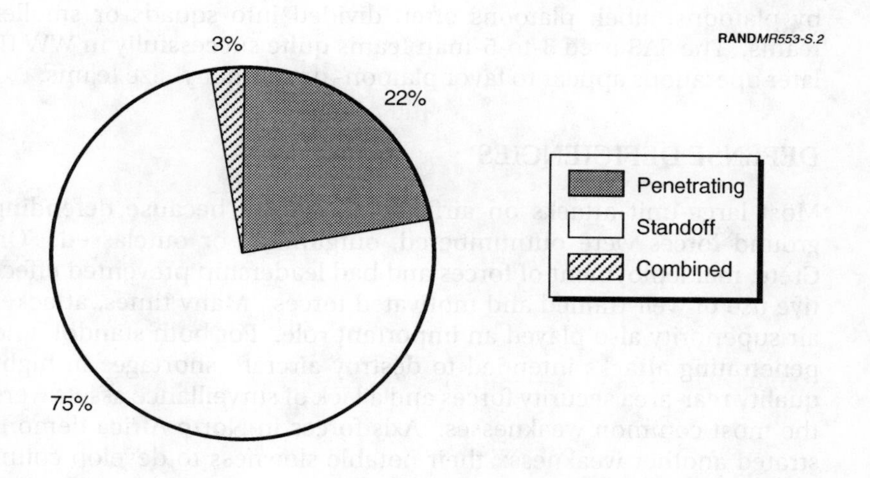

Figure S.2—Air Base Attack Tactics, 1940–1992

North Vietnam Army (NVA) attackers rarely used penetrating tactics, relying on standoff weapons for 96 percent of their attacks. Recent attacks have used both techniques. Kurdish and Filipino insurgents used penetrating tactics; insurgents in El Salvador and Afghanistan used standoff weapons. The Special Air Services (SAS) attack against the Argentine airstrip on Pebble Island used both techniques, opening the attack with naval gunfire and light antitank weapons, then moving onto the airfield to plant charges on aircraft. It is likely that both tactics will continue to be used in the future, depending on the quality of perimeter defenses. Where perimeter defenses are weak, attackers will probably continue to penetrate and place charges. More troublesome is the possibility that precision-guided munitions for both existing standoff weapons and some new weapons may give small standoff attacks a lethality they lacked in the past.

Large forces are not required to conduct the most common type of air base attack. As one would expect, attempts to capture airfields or to deny their use have required larger forces, typically of regimental strength. In contrast, quite small forces have been used in efforts to destroy aircraft and equipment. Such attacks are typically conducted by platoons, albeit platoons often divided into squads or smaller teams. The SAS used 3-to-5-man teams quite successfully in WW II; later operations appear to favor platoon- or company-size teams.

DEFENSE DEFICIENCIES

Most large-unit attacks on airfields succeeded because defending ground forces were outnumbered, outgunned, or outclassed. On Crete, maldeployment of forces and bad leadership prevented effective use of well-trained and motivated forces. Many times, attacker air superiority also played an important role. For both standoff and penetrating attacks intended to destroy aircraft, shortages in high-quality rear-area security forces and a lack of surveillance assets were the most common weaknesses. Axis forces in North Africa demonstrated another weakness: their notable slowness to develop countermeasures to SAS attacks. In particular, their failure to establish night listening posts and ambushes outside of airfield perimeters is perplexing; such practices would not have taken large forces and could have paid large dividends. Conversely, U.S. forces in Vietnam demonstrated great innovation and creativity in their defensive countermeasures. Joint-force responses to penetration attacks proved quite effective. Military Assistance Command, Vietnam's

(MACV's) refusal to make air base defense a high priority for such resources as ground forces and airborne surveillance assets, however, made it impossible to counter the standoff threat effectively. Without ground forces and airborne surveillance assets dedicated to controlling the standoff footprint,[5] USAF bases remained vulnerable to the end of the war.

Reliance on other services for the defense of air bases was a problem for the RAF on Crete, the *Luftwaffe* in North Africa, and the USAF in Vietnam. In each case, air base defense had to compete with other missions to which ground commanders assigned higher priority. On Crete, ground commanders failed to recognize that air bases were key terrain that the attacker must be denied at all costs. In North Africa, *Luftwaffe* units reported up their own chain of command and were not integrated under General Rommel, the theater commander, which hampered the coordination of defenses.

STRATEGIC EFFECT OF THE ATTACKS

What effect have these attacks had on the outcome of the subject conflicts? At the least, they caused the loss of valuable aircraft, materiel, and personnel, and they forced the defenders to devote substantial resources to the defense of their airfields.

In one case—British special forces' attacks on Axis airfields in North Africa—the loss of aircraft was so severe and the airpower balance so precarious that these small attacks made a major contribution to the RAF's battle against the *Luftwaffe*. In others, the loss of airfields to attacking forces enabled the attacker's air force to move in and extend its range. In the Pacific theater, the need to capture and defend airfields drove both American and Japanese campaign planning. For example, the Japanese victory over the British in Malaya was made possible when critical air bases were captured by ground forces. The U.S. island-hopping campaign was focused on capturing airfields; toward the end of WW II, Tinian, Okinawa, and Ie Shima were captured to launch air attacks against the Japanese homeland. The

[5]The *standoff footprint* is the area around a base from which weapons can be fired onto aircraft and other targets. Its size varies with the type of weapon; typically, it extends 10 kilometers beyond the perimeter fence.

Japanese attack on Midway sought to capture the island for its airfield; their failure to do so and their losses incurred in the process marked a turning point in the war.

CONCLUSIONS

It is clear from this analysis that ground attacks on airfields in past conflicts cannot be dismissed as a quaint subfield of military history. The basic techniques of airfield attack and defense have not changed dramatically over the past 50 years. The simple-but-effective tactics and the strategic rationale for the attacks are as relevant today as they were in 1940. Indeed, the centrality of airpower to modern warfare makes airfields even more tempting targets than they have been. Conversely, a variety of new information and sensor and weapon technologies offers opportunities for attacker and defender alike. It remains to be seen who will exploit these opportunities most effectively.

What lessons can be learned from this historical review? The five primary conclusions of this study are as follows:

- The most common air base attack objective was to destroy aircraft.
- Seventy-five percent of the 645 attacks used standoff weapons.
- Standoff attacks have proved extremely difficult to counter.
- Reliance on non–air force services for air base defense proved problematic for Britain's Royal Air Force (RAF) on Crete, the German *Luftwaffe* in North Africa, and the USAF in Vietnam.
- Small forces using unsophisticated weapons have successfully destroyed or damaged over 2,000 aircraft.

During World War II, ground forces attacked air bases in pursuit of three of the four objectives (harassment not included). During the Vietnam War, virtually all air base attacks were focused on only two objectives: destroy aircraft and harass defenders. Of the 19 attacks since Vietnam, 12 have sought to destroy aircraft. Of the remaining 7, 5—by the Soviet Union in Afghanistan and the United States in Grenada and Panama—were to capture airfields for use as airheads and may not be representative, because few other nations have this capability. To the extent that we wish to look to historical experience as a predictor of future challenges, these cases are probably misleading.

It is highly unlikely that USAF bases will be assaulted by large airborne forces in the near future. Although the possibility of large-unit attacks on airfields should not be discounted, it is a possibility more for adversaries of the United States than for the United States: The United States has elite airborne units that specialize in assaulting and capturing airfields. Airborne insertion of special forces is another matter, however, and a distinct possibility in a future Korean conflict, for instance. The threat facing USAF bases in future contingencies is more likely to resemble that presented by SAS operations in WW II or the VC/NVA operations in Vietnam. If the historical experience is any indication, standoff threats will continue to pose a particularly daunting challenge. New precision-guided munitions for mortars and other standoff weapons will only exacerbate this problem.

In conclusion, attacks by small forces with the limited objective of destroying aircraft succeeded in destroying or damaging over 2,000 aircraft between 1940 and 1992. Such attacks are powerful testimony to the effectiveness of small units against typical air base defenses and offer a sobering precedent for those responsible for defending USAF bases against them.

ACKNOWLEDGMENTS

This study benefited greatly from the assistance of many people in the Air Force, at RAND, and elsewhere.

In the Air Force, Brigadier General Stephen Mannell, Chief of Security Police, has provided tremendous support to the air-base-defense study from its inception and has opened doors throughout the Security Police community. His interest in the historical work is particularly appreciated. Colonel Steve Shoemaker and Lieutenant Colonel Robert Tirevold, both on General Mannell's staff, were our first contacts in the Security Police community and acted as tutors, constructive critics, and colleagues. They went out of their way to be helpful and did the legwork that made all the field trips possible. Other Security Police officers in the field were gracious hosts and shared their experiences and extensive knowledge of air-base-defense operations. Those who helped specifically with this historical work include Colonel Michael Rader, Colonel Jack Costello, Colonel Rocky Lane, and Colonel Richard Larkins.

Several historians in the Air Force Center for History (formerly the Office of History) made helpful suggestions and provided access to unit histories, Project CHECO reports, and other useful documents relating to air base attacks in Vietnam. I especially want to thank Dr. Richard Hallion, the Air Force Historian; Dr. Wayne Thompson; Dr. Shelton Goldberg; and Lieutenant Colonel Al Howey. Mr. Archie DiFante oversaw a remarkably fast declassification review of the Project CHECO report on base defense in Thailand.

Squadron Leader Nick Tucker, Deputy Officer in Charge of the Training Wing, Royal Air Force Regiment Depot at Honington,

England, kindly allowed me access to his manuscript on regimental combat awards.

John Kreis, at the Institute for Defense Analysis, made several helpful suggestions on the draft report.

At RAND, Bob Howe and Mark Lorell provided thorough and thoughtful technical reviews. David Adamson's advice on organization and structure is much appreciated; Marian Branch's editing greatly improved the manuscript. I also thank Ann Flanagan for her assistance in analyzing the Vietnam data. Debbie Peetz, the Washington Office librarian, tracked down and ordered often-obscure historical references. Sandy Petitjean worked her usual wonders with the graphics. Finally, I want to acknowledge the helpful comments of colleagues Chris Bowie, Nicole DeHoratius, David Shlapak, Lieutenant Colonel Mitchell Sivas, and Ken Watman.

Chapter One

INTRODUCTION

Recent RAND research on trends in global airpower suggests that few opponents will be able to challenge the United States Air Force (USAF) in the air. If that is correct, future adversaries are likely to look for alternative means to counter U.S. airpower. A RAND study for the Air Force has been investigating means that future adversaries might pursue to counter U.S. airpower.

This historical effort was part of that study and sought to better understand past, present, and future ground threats to air bases. In the course of this research, it became clear that attacks on air bases were much more frequent and successful than is commonly appreciated. For this reason, the history of those attacks is pertinent to future USAF operations. As the reader will see, the basic techniques of airfield attack and defense have not changed markedly over the past 50 years. Conversely, a variety of new information, and sensor and weapon technologies offer opportunities for attacker and defender alike. It remains to be seen who will exploit these opportunities most effectively.

This report presents a comprehensive overview of ground attacks on air bases from the first known attacks in 1940 to the most recent in 1992. The objectives, tactics, and outcomes of those attacks are analyzed to identify lessons learned that can be applied to future conflicts. In particular, this report identifies the attack techniques that proved most difficult to counter and offers some suggestions for improving air base defenses against them.

BACKGROUND

In 1921, Italian Army General Giulio Douhet observed that "it is easier and more effective to destroy the enemy's aerial power by destroying his nests and eggs on the ground than to hunt his flying birds in the air."[1] Douhet's metaphor was directed at fellow airmen, pointing out both the great offensive potential of airpower—a radical notion in 1921—and the exceptional vulnerability of aircraft on the ground. Flying machines, even modern ones, by their very nature are thin-skinned, relatively soft targets. Speed, maneuverability, and stealth enable these unarmored vehicles to survive and be decisive in combat. These characteristics are absent, in contrast, when an aircraft is parked on a ramp. Furthermore, compared with most other ground targets, a parked aircraft is triflingly easy to destroy. During World War II (WW II), attacks on airfields were common and highly successful. The 1967 Israeli raid on Egyptian airfields reminded the world's air forces that unsheltered aircraft remain a tempting target in modern air warfare. Since 1967, billions of dollars have been spent on aircraft shelters, air defenses, and programs to enhance air base survivability.

Douhet's observation, like most great insights, has applicability beyond the confines of its initial setting. If aircraft are vulnerable on the ground, why not attack them with every weapon available? That is just what the world's armies have done at least 645 times[2] in ten conflicts between 1940 and 1992, destroying or damaging over 2,000 aircraft. Ground attack forces have included airborne, airmobile, infantry, and armor elements. Airborne forces have arrived on the objective by parachute, glider, and aircraft landing, often under fire. Armor and infantry have assaulted by land, and amphibious forces have landed by sea. More recently, helicopters have been used to transport the assault force. Finally, special forces, guerrillas, and terrorists have made their contribution.

---

[1]Giulio Douhet, *The Command of the Air*, Washington, D.C.: Office of Air Force History, 1983 (originally published in 1921), pp. 53–54.

[2]This number is based on *deliberate* attacks on airfields, whether they were independent operations or part of a larger offensive. It does not include the many times that ground forces overran airfields on their way to other objectives.

As Figure 1.1 shows, these attacks have occurred in quite diverse locations around the world. The most recent attacks—El Salvador in 1990, Puerto Rico in 1991, Iraq in 1991, and the Philippines in 1992—span the globe.

What effect have these attacks had on the outcome of their subject conflicts? At the least, they caused the loss of valuable aircraft, materiel, and personnel, and forced the defenders to devote substantial resources to the defense of their airfields. In one case—British special forces' attacks on Axis airfields in North Africa—the loss of aircraft was so severe and the airpower balance so precarious that these attacks appear to have made a major contribution to the British cause. In others, the loss of airfields to attacking forces enabled the attacker's air force to move in and extend its range. For example, the Japanese victory over the British in Malaya was made possible by ground forces' capturing critical air bases. Finally, in both North

RANDMR553-1.1

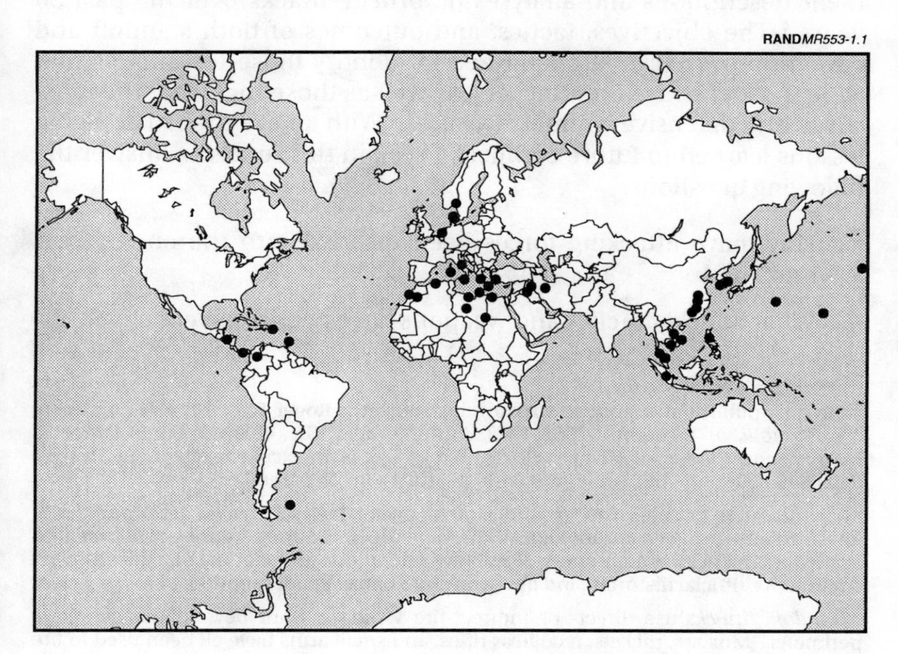

Figure 1.1—Locations of Ground Attacks on Airfields, 1940–1992

Africa and the Pacific theater, the need to capture and defend airfields drove Axis and Allied campaign planning. It is clear from this analysis that ground attacks on airfields in past conflicts cannot be dismissed as a quaint subfield of military history. The simple-but-effective tactics and the strategic rationale for the attacks are as relevant today as they were in 1940. Indeed, the centrality of airpower to modern warfare makes airfields even more tempting targets than in the past.

PURPOSE

Given the numerous occurrences, global distribution, and recentness of ground attacks on air bases, it is surprising that a comprehensive history of those attacks has not been published.[3] This report is intended to begin filling that void by bringing together in one document descriptions and analyses of airfield attacks over the past 50 years.[4] The objectives, tactics, and outcomes of both standoff and penetrating attacks[5] are analyzed to identify those techniques that were most effective for attackers, as well as those that were the most successful defensive countermeasures. With an eye toward applying lessons learned to future conflicts, I seek in this report to answer the following questions:

- How have attacking forces been inserted into the enemy rear area?
- What attacker tactics and weapons have been most effective?

---

[3]The only published historical work on this subject is Roger Fox, *Air Base Defense in the Republic of Vietnam: 1961–1973*, Washington, D.C.: U.S. Air Force Office of History [now Center for History], 1979. Fox's book is an excellent history of air base defense in Vietnam, but it does not address attacks in other conflicts.

[4]This research revealed 645 ground attacks on air bases between 1940 and 1992. These attacks are listed chronologically in Appendix B. Sources for the entries include academic and personal accounts of military operations and campaigns, government documents, official histories, and newswire and/or newspaper reports.

[5]*Standoff attacks* use direct- or indirect-fire weapons from beyond the defensive perimeter. Mortars, rockets, recoilless rifles, and small arms have all been used to fire on aircraft, facilities, and personnel from distances up to 11 kilometers. *Penetrating attacks* typically are done covertly by small teams who slip through the defensive perimeter and place bombs with time fuzes (*satchel charges*) on aircraft and materiel. Defensive perimeters have been assaulted outright also; the attackers then use direct-fire weapons (machine guns, tank guns, and small arms) against airfield targets.

- What defensive countermeasures have worked?
- Were there promising countermeasures that the defense failed to employ?
- What has been the strategic effect of ground attacks on air bases in previous conflicts?

ORGANIZATION

This report is organized as follows. Chapter Two identifies four objectives of air base attackers and gives historical examples for each. Since the attacker's force size tracked closely with the ambitiousness of its objectives, this classification system can help defenders make the conceptual link between threats and appropriate defensive organizations, tactics, and weapons. For example, theater-level offensives that used large ground formations to capture airfields could not be countered solely by USAF Security Police units; offensives at this level are the responsibility of the theater commander. As is pointed out throughout the report, the most common challenge for past air base defenders has been to detect and stop *small forces* seeking to destroy aircraft.

Chapters Three through Five present short case studies of air base attacks on Crete, in North Africa, and in Vietnam. The cases selected represent both historical and current threats to air bases, have a considerable literature or body of data available, and offer insights into insertion techniques and attack tactics and weapons that continue to be used by air base attackers.

Chapter Three discusses the successful May 1941 German airborne assault of Commonwealth airfields on Crete. The loss of Crete convinced the British government that a dedicated air-base-defense force was needed and led to the creation of the Royal Air Force Regiment. USAF leaders followed suit, creating a dedicated air-base-defense force also. Thus, the attack on Crete is a seminal event for the Royal Air Force Regiment and a touchstone for U.S. Air Force Security Police.

This episode is relevant for two additional reasons. First, there is a great deal of confusion in air-base-defense circles about what really happened on Crete. This chapter seeks to dispel several myths about the operation and to offer lessons that are substantiated by the excellent historical materials available. Second, the British experience on Crete illustrates problems associated with joint-service operations in defense of airfields.

Chapter Four investigates British special forces' attacks on Axis airfields in North Africa and the Mediterranean between 1940 and 1943. It is included for two reasons. First, the British were highly successful, destroying almost 400 aircraft in 50 raids. Second, the extensive literature on British special forces—both first-person accounts and scholarly works—offers an attacker's perspective on the purpose, challenges, and tactics of air base attack.

Chapter Five analyzes Viet Cong (VC) and North Vietnamese Army (NVA) attacks on U.S. air bases in Vietnam and Thailand. Vietnam was included for several reasons. The sheer number of attacks—475 against main operating bases (MOBs) alone—argues for inclusion, because the Vietnam experience dominates the historical record. Second, the VC and NVA demonstrated impressive creativity, adaptability, and persistence in their operations. They proved that third-world forces with relatively crude weapons can threaten a superpower's airfields. Excellent mission planning, intelligence preparation, training, and leadership made up for what they lacked in equipment. Finally, the USAF kept excellent records on air base attacks in Vietnam. Project CHECO[6] reports and the official USAF history of air base defense in Vietnam[7] offer detailed insights into what organizational structures, training, equipment, and tactics were most effective. Of particular interest is a recently declassified Project CHECO report on five attacks on USAF air bases in Thailand.[8] Chapter Five analyzes data on these five attacks to gain additional insights into the air-base-defense problem in Vietnam.

---

[6]Contemporary Historical Examination of Current Operations, an effort by Pacific Air Forces' Headquarters Staff to identify lessons learned from operations during the Vietnam conflict.

[7]Fox, 1979.

[8]USAF, *Base Defense in Thailand: Project CHECO Southeast Asia Report,* Hickam AFB, Hawaii: Headquarters, Pacific Air Forces, February 18, 1973. (Declassified by USAF Office of History, August 16, 1994.)

Chapter Six presents lessons learned and conclusions for future conflicts.

Following the successful May 1941 German attack on Crete, British Prime Minister Winston Churchill issued a directive on improving air base defense. Contemporary discussions of Crete have often cited parts of this memo out of context. To give the reader a better understanding of Churchill's concerns, I give the full text of the memo in Appendix A. Appendix B includes short descriptions of the 645 attacks identified in the course of this research. Full citations for all references in Appendix B can be found in the Bibliography.

Chapter Two

HISTORICAL OVERVIEW

A RANGE OF THREATS

Air bases have been attacked as a way of pursuing a broad range of objectives, from the ambitious goal of capturing an airfield to the minimalist goal of harassing air base operations. Discussions of air base defense often treat these threats as similar. However, the threats are quite different and call for a broad range of defensive countermeasures. To bound the problem and have the range and nature of historical threats to air bases become more visible, I categorize the attacks identified in this research according to the attacker's major objective. These attacks can be grouped into the following four broad categories (the number of each attack type follows each objective):

- Capture airfield (41)
- Deny defender use of airfield (47)
- Harass defenders (173)
- Destroy aircraft and equipment (384).

As Figure 2.1 illustrates, the majority of these attacks sought to destroy aircraft and equipment. Only 6 percent of the attacks attempted to capture air bases to insert troops or so that the attacker could carry out its own offensive air operations. Most of the larger attacks occurred during World War II, although Soviet forces in Afghanistan (1979) and U.S. forces in Grenada (1983) and Panama (1989) also seized airfields for use as airheads.

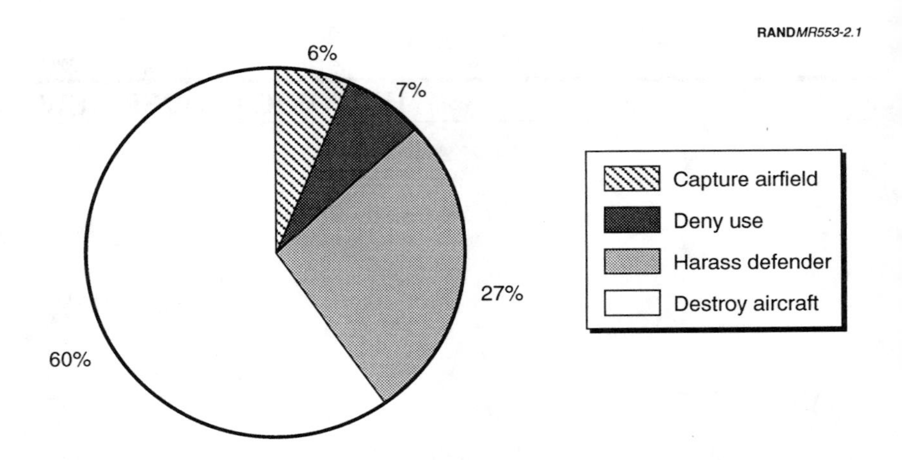

Figure 2.1—Airfield-Attack Objectives

This chapter presents and discusses historical examples of attacks that pursued these four objectives. The discussion is largely descriptive and is intended to briefly introduce air base attack and to provide a context for the analysis that follows. Additional details on these attacks can be found in Appendix B.

OBJECTIVE I: CAPTURE AIRFIELD

Ground forces have sought to capture airfields on 41 occasions. In 16 cases, airborne forces attacked airfields to use them as airheads for the insertion of additional troops. In 23 cases, airfields were attacked so that the aggressor's air force could use the facilities. In 2 cases, the airfields were attacked to destroy collocated ground forces.[1] The two main goals are detailed below.

---

[1]On February 27, 1991, armored units from the U.S. 24th Infantry Division (Mechanized) captured Iraqi airfields at Jalibah and Talil, destroying 29 aircraft in the process. See Appendix B for more details.

Seize Airfield as Airhead

Airborne forces attacked airfields 16 times between 1940 and 1989 to secure airfield facilities for the air landing of troops, heavy equipment, and supplies.

The Germans were the first to recognize the value of adversary airfields as insertion points for their own forces. On April 9, 1940, German paratroopers captured airfields at Aalborg, Denmark, and Sola, Oslo, and Stavanger, Norway. At Oslo, after the initial airborne assault was aborted because of heavy anti-aircraft fire, a few aircraft landed under fire and discharged paratroopers, who captured the airfield. On May 10, 1940, German paratroopers captured three airfields at The Hague and another at Rotterdam in Holland. At The Hague, Dutch reserve forces drove the Germans off the airfields. German ground forces relieved the paratroopers five days later, recapturing the airfields.[2] A year later, in the largest German airborne operation of the war, German paratroopers assaulted the three Commonwealth airfields on Crete. Driven off by stout defenses at two airfields, the Germans did, however, capture the field at Maleme. Using Maleme as an airhead, the *Luftwaffe* rapidly reinforced the tenuous German toehold. Crete fell to the Germans a week later.

In 1979, Soviet airborne forces seized Kabul airport and several other bases for use by follow-on forces in the Soviet takeover of Afghanistan.[3] In 1983, American forces conducted similar attacks in Grenada during *Operation Urgent Fury*, capturing Salinas and Pearls airports on Grenada by airborne and helicopter assault, respec-

---

[2]Donald E. Cluxton, Jr., "Concepts of Airborne Warfare in WWII," Master's Thesis, Duke University, Durham, N.C., 1967, pp. x, xi, xvii; Thomas E. Greiss, ed., *The Second World War: Europe and the Mediterranean*, The West Point Military History Series, Wayne, N.J.: Avery Publishing Group, Inc., 1984b, p. 29; Thomas E. Greiss, ed., *Atlas for the Second World War: Europe and the Mediterranean*, The West Point Military History Series, Wayne, N.J.: Avery Publishing Group, Inc., 1985b, Map 8a.

[3]Drew Middleton, "Soviet Display of Flexibility: Afghan Airlift Is Lesson in Moving Troops Fast," *The New York Times*, December 28, 1979, pp. A1, A13.

tively.[4] Finally, in 1989, during *Operation Just Cause,* U.S. Rangers captured Rio Hato and Tocumen airfields in Panama.[5]

Capture Airfield for Offensive Air Operations

In 23 cases, attacking air forces sought to capture enemy airfields to perform their own air operations. This objective appears to be exclusive to World War II; no cases were found in other conflicts. In many of the 23 cases, the attackers were able to mount offensive air operations within hours or days after their ground forces had secured the airfield. (Ground forces captured the airfields so that their own air force could fly in and conduct offensive air operations from the airfield, extending the reach of the attacker's air force.)

Fighting in the Pacific theater was noted for its jointness, which integrated ground, naval, and air operations to an unprecedented degree. In particular, the campaign plans of both sides were largely determined by the need to capture and defend air bases.[6] Thus, joint offensive operations were often launched to capture enemy airfields. Subsequent air operations from these new bases extended the offensive reach of airpower, allowing for new naval and ground operations that seized new airfields.

In December 1941, Japanese ground and naval forces attacked Wake Island for its airfield. They were initially beaten off but returned several weeks later and captured the island. Also in December, Japanese forces invaded Thailand and Malaya. Their ultimate objective was to defeat British forces in Malaya and capture Singapore. An important intermediate objective was to defeat the Royal Air Force (RAF). To do so, the Japanese Air Force needed air bases in Thailand and northern Malaya. The Japanese 25th Army made amphibious landings at Singora and Patani, Thailand, and Kota Bharu, Malaya. In a week of fighting, they captured the Thai airfields at Singora and Patani, and RAF bases at Kota Bharu, Alor Star, and Sungei Patani, Malaya.

---

[4]Mark Adkin, *Urgent Fury: The Battle for Grenada,* Lexington, Mass.: Lexington Books, 1989, pp. 200, 214, 217, 236.

[5]Malcolm McConnell, *Just Cause: The Real Story of America's High-Tech Invasion of Panama,* New York: St. Martin's Press, 1991, pp. 73, 99, 191.

[6]I am indebted to RAND colleague Bob Howe for this observation.

Japanese aircraft used these bases to attack RAF installations throughout Malaya. British forces in Singapore surrendered to the Japanese on February 15, 1942.[7]

Other examples of this type of attack include the

- capture of the RAF airfield at Palembang, Sumatra, in February 1942[8]
- Japanese attack on Midway Island in June 1942[9]
- British assault on the Vichy French airfield at Souk-el-Arba, Algeria, in November 1942[10]
- U.S. landings on Tinian, Iwo Jima, Okinawa, and Ie Shima in 1944 and 1945.

OBJECTIVE II: DENY ENEMY USE OF AIRFIELDS

In 47 cases, the attacker sought to counter the defender's airpower by capturing or shutting down operations at air bases.

Four of the cases were from *Operation Torch*, the November 1942 Allied invasion of Algeria. Fearing that Vichy French aircraft might intercept Allied transports during the initial days of the invasion, Allied planners assigned airborne forces to capture French airfields at La Senia, Duzerville, and Youks-les-Bains, Algeria. Bad weather

---

[7]John F. Kreis, *Air Warfare and Air Base Air Defense,* Washington, D.C.: Office of Air Force History, 1988, p. 101; Thomas E. Greiss, ed., *Atlas for the Second World War: Asia and the Pacific,* The West Point Military History Series, Wayne, N.J.: Avery Publishing Group, Inc., 1985a, Map 7; B. H. Liddell Hart, *History of the Second World War,* New York: G. P. Putnam's Sons, 1970, p. 225.

[8]Nick Tucker, "In Adversity: Exploits of Gallantry and Awards in the RAF Regiment and Its Associated Forces," unpublished manuscript.

[9]U.S. aircraft discovered the Japanese fleet and sank four carriers, turning back the invasion force before it could land. RAND analyst David Shlapak has observed that this may be the most strategically significant airfield attack in history. The resulting battle, and the destruction of the core of the Japanese attack force, was a turning point in the war. For more on the Battle of Midway, see Thomas E. Greiss, ed., *The Second World War: Asia and the Pacific,* The West Point Military History Series, Wayne, N.J.: Avery Publishing Group, Inc., 1984a, pp. 111–115.

[10]Wesley F. Craven and James L. Cate, eds., *The Army Air Forces in World War II, Vol. II,* Chicago, Ill.: University of Chicago Press, 1949, p. 81.

caused the cancellation of the airborne assault on La Senia but did not prevent an armored column from reaching and capturing the airfield on November 12. The same day, British paratroopers captured Duzerville airdrome. Three days later, U.S. paratroopers captured Youks-les-Bains.[11]

Four additional cases were from *Operation Ichigo*, the Japanese theater offensive launched to capture General Claire Chennault's 14th Air Force bases in East China. Chennault's force had so disrupted Japanese logistics that the commander of Japan's North China Area Army felt compelled to launch a ground campaign to seize the air bases. Between September and November 1944, Japanese ground forces captured Chennault's bases at Ling Ling, Tanchuk, Kweilin, and Liuchow, China.[12]

Thirty incidents are associated with Japanese attacks on the British airfield at Meiktila, Burma, in March 1945. Virtually every night during that month, Japanese forces made multiple attempts to capture the airfield. Each night, the Royal Air Force pulled its aircraft into small perimeters defended by Royal Air Force Regiment and other Commonwealth ground forces. Every night the Japanese attacks were beaten back, and every dawn the airfield would be cleared of any remaining Japanese soldiers and flight operations would resume.[13]

Finally, after the United Nations (UN) landing at Inchon, Republic of North Korea, the U.S. Air Force tried to use the sod landing strip at Kunsan, but harassment from North Korean guerrillas prevented such use for several months.[14]

---

[11]Craven and Cate, *Vol. II*, 1949, pp. 68–81.

[12]Craven and Cate, *Vol. V*, 1953, pp. 220–225; Charles F. Romanus and Riley Sunderland, *United States Army in World War II, China-Burma-India Theater: Stilwell's Command Problems*, Washington, D.C.: Department of the Army, 1956, pp. 316–328, 405–408.

[13]Tucker, unpublished.

[14]Lawrence V. Schuetta, *Guerrilla Warfare and Airpower in Korea, 1950–53*, Maxwell AFB, Ala.: Aerospace Studies Institute, 1964, p. 38.

OBJECTIVE III: HARASS DEFENDERS

Enemy forces seeking to realize Objectives I, II, and IV certainly harassed defenders and disrupted base operations. The purpose of a separate category is to identify those attacks whose primary objective was harassment. All but one example of this objective are from the Vietnam War.

The Viet Cong and NVA conducted 448 standoff attacks against allied air bases, 172 of which fired fewer than five rounds and did no damage to aircraft. Such attacks appear not to have been serious attempts to destroy aircraft. The strategic purpose of air base attacks in general was to kill Americans, cause damage, and demonstrate allied vulnerability—the ultimate objective being to undermine U.S. popular support for the war. These outcomes suggest that harassment was the primary purpose of smaller standoff attacks. Damage to aircraft was a bonus for such attacks, but not central to mission accomplishment. Conversely, the attacks that did the most damage were carefully planned, were given the necessary manpower and materiel, and were executed to maximize damage to aircraft and equipment. Such attacks are counted against Objective IV.

The one incident outside of Vietnam is from *Just Cause,* the 1989 American intervention in Panama. About the time that the U.S. operations began, unknown gunmen fired small arms on a hangar at Albrook Air Station, Panama. The attackers may have hoped to get lucky and cause damage to aircraft, but their force was small and the attack was brief. These circumstances, combined with the fact that the attackers fired from outside the airfield fence, suggest that this incident belongs in the harassment category.[15]

OBJECTIVE IV: DESTROY AIRCRAFT AND EQUIPMENT

Sixty percent of the attacks discussed in this report sought to destroy aircraft and equipment. Although aircraft and equipment were almost certainly damaged or destroyed in attacks pursuing Objectives I through III, few records were found listing aircraft losses for those operations.

---

[15]McConnell, 1991, p. 112.

The first recorded attempt to destroy parked aircraft with ground forces was in October 1940, when British special forces destroyed an Italian bomber in North Africa. Over the next two years, these small teams, operating hundreds of miles behind enemy lines, destroyed 367 Axis aircraft.

During the Korean War, North Korean infantry attempted to penetrate the perimeter of Pohang airfield in South Korea. They were stopped by an ad hoc ground defense force composed of Security Police, mechanics, and other support personnel. This was a short-lived victory, because the airfield was evacuated a few days later, when the nearby port city of Pohang fell to North Korean regular forces.[16]

The Vietnam War is responsible for 316 of the attacks in this category (82 percent of the total). These attacks destroyed 393 U.S. and allied aircraft and damaged another 1,185. The most common attack technique was for a small team to fire ten or fewer mortar, rocket, or recoilless rifle rounds at an air base, then flee. Conversely, only 21 sapper[17] attacks were made against air bases in Vietnam and Thailand. An additional eight attacks combined sapper and standoff techniques.

In a well-planned and -executed operation, an unknown number of Puerto Rican nationalists, the *Macheteros*, slipped through a hole in the fence at Muñiz Air National Guard Base (San Juan, Puerto Rico) on January 12, 1981, and planted satchel charges on 11 aircraft. The *Macheteros* then escaped without detection. Sixty minutes later, the charges blew, destroying eight A-7D aircraft and damaging two. Two A-7s escaped damage because the satchel charges placed on them were duds. One non-operational F-104 aircraft on display was also destroyed.[18] (See photo plates.)

---

[16]Robert F. Futrell, *The United States Air Force in Korea, 1950–1953,* Washington, D.C.: U.S. Air Force Office of History, 1983, p. 124.

[17]Strictly speaking, *sappers* are military engineers who specialize in constructing field fortifications or laying minefields. During the Vietnam War, the term was widely used to describe enemy infantry who penetrated base defenses to place explosive charges.

[18]Jo Thomas, "Puerto Rico Group Says It Struck Jets," *The New York Times,* January 13, 1981, pp. A1, A12.

A year later, on January 27, 1982, Faribundo Marti National Liberation Front (FMLN) guerrillas attacked Illopango AFB in El Salvador. Using rockets and sappers, the FMLN destroyed five helicopters, five fighter aircraft, and five transport aircraft. An additional seven aircraft were damaged.[19]

In 1982, during the Falklands War, the British Special Air Service (SAS) was tasked to raid an Argentine airstrip. British commanders feared that the Argentinians would use light attack aircraft based on Pebble Island to attack British ground forces during the upcoming amphibious assault in San Carlos Bay, 30 miles to the southeast. The SAS mission was to destroy those light aircraft.

Before sunrise on May 15, a 45-man SAS detachment was inserted onto Pebble Island by helicopter. The men walked the final 6 kilometers to the airstrip, then assaulted, firing small arms and 66-mm rockets at the Argentine aircraft while naval gunfire from HMS *Glamorgan* provided suppressive fire. Elements of the force then went onto the airstrip, planting charges on the aircraft.

Ten light attack aircraft were damaged or destroyed, along with one transport. A ton of ammunition and a radar station were also destroyed, and naval gunfire badly cratered the airstrip. Although several of the aircraft could have been repaired, the Argentinians lacked the facilities on Pebble Island to do so. The airstrip was out of action for the remainder of the conflict. After suffering through several British air raids, Argentine helicopters evacuated the last personnel from Pebble Island on May 31.[20]

In 1986, Afghan guerrillas struck twice against Soviet forces at Shindad Air Base in Afghanistan. On May 27, they used a SAM-7 man-portable (surface-to-air) missile to shoot down a Soviet transport on approach, and, on May 30, they launched a 25-minute-long standoff attack. Using 60 107-mm rockets, they destroyed two jet

---

[19]"Guerrilla Attacks Intensify," *Facts on File*, February 5, 1982.

[20]Jeffrey Ethell and Alfred Price, *Air War South Atlantic*, New York: Macmillan Publishing Co., 1983, pp. 65–66; Max Hastings and Simon Jenkins, *Battle for the Falklands*, New York: W. W. Norton, 1983, pp. 186–187; Rodney A. Burden et al., *Falklands: The Air War*, Dorset, England: Arms and Armour Press, 1986; John Strawson, *A History of the S.A.S. Regiment*, London: Secker and Warburg, 1984, pp. 231–232.

fighters and six helicopters. A large fuel tank was also hit, and it burned for two days.[21]

Three years later, during *Operation Just Cause*, a 54-man detachment of U.S. Navy SEALs (sea/air/land personnel) destroyed Manuel Noriega's Learjet at Paitilla airport in Panama.[22] Also in 1989, unknown attackers firebombed a U.S. Department of State aircraft at Monteria, Colombia. The aircraft supported Colombian government anti-drug operations.[23]

In 1990, FMLN guerrillas again attacked an El Salvadoran air base, damaging one aircraft.[24] The year 1991 saw two attacks against airfields in Puerto Rico and Iraq. On March 17, terrorists struck Muñiz Airport, setting fire to one A-7 aircraft and causing $100,000 in damage.[25] On March 28, Kurdish insurgent sappers penetrated the defenses of Khalid Air Base in Iraq, destroying three Su-22 jet fighters in hardened shelters and four MI-8 helicopters on the ramp.[26] The most recent attack on an air base occurred on November 5, 1992, when 100 guerrillas attacked a Philippine air force base in the northern province of Isabela, destroying two OV-10 Bronco aircraft and damaging a Sikorsky S-76 helicopter.[27]

---

[21]Barry Renfrew, *Guerrillas Report Attack on Major Soviet Air Base,* Associated Press Report, dateline: Islamabad, Pakistan, June 8, 1986.

[22]McConnell, 1991, pp. 47–72, 219.

[23]RAND Terrorism Database. RAND'S Terrorism Database contains 75 terrorist attacks on individuals or installations associated with various nations' air forces. The three attacks included here were the only ones that appeared to be serious attempts to destroy aircraft.

[24]Associated Press, "Salvadoran Rebels Hit Military Posts," *Chicago Tribune,* November 21, 1990, p. 3.

[25]"Intruders Damage Plane at U.S. Base in Puerto Rico," *Los Angeles Times,* March 18, 1991, p. A15.

[26]United Press International, "Kurdish Guerrillas Attack Air Base, Destroy Aircraft," dateline: Athens, Greece, March 28, 1991.

[27]Agence France Presse, "Communist Guerrillas Destroy Two Air Force Planes," dateline: Manila, Philippines, November 6, 1992.

ATTACK BREAKDOWN BY CONFLICT

Table 2.1 breaks down these incidents by conflict. Seventy-six percent of the recorded incidents occurred during the Vietnam War; 20 percent occurred in World War II; and the remaining 4 percent took place in 8 other modern conflicts and terrorist attacks. Note that the World War II numbers are probably underestimated. Unlike the United States in the Vietnam War, World War II combatants did not keep detailed records of tactical engagements at the hundreds of airfields around the world.[28] In several cases included in this report, multiple attacks were reported as one incident because it was impossible to determine the exact number of attacks.[29]

The information on the 130 WW II attacks discussed in this report was derived from personal accounts, official histories, and general accounts of campaigns. For most military historians, ground action against airfields was not of sufficient interest to warrant the effort

Table 2.1

Ground Attacks on Airfields

Conflict	Number of Incidents	Aircraft Destroyed/ Damaged
World War II	130	367/NA
Korea	3	0
Vietnam	493	393/1,185
Falklands	1	11
El Salvador	2	15/18
Grenada	2	0
Afghanistan	3	9
Panama	4	1
1991 Gulf War	3	36
Philippines	1	2/1
Terrorism	3	9/3
TOTAL	645	843/1,207

NA = data are not available.

[28]These airstrips were often nothing more than a grass or dirt strip with a few tents or simple structures.

[29]For example, between August 1942 and February 1943, Japanese forces attacked Henderson Field on Guadalcanal many times.

necessary to collect the data. Furthermore, the World War II aircraft losses discussed in this report are all from small-unit action—primarily that of British special forces. Larger-scale ground attacks on airfields in Europe and Asia probably accounted for a fair number of aircraft losses, but reports on those attacks do not provide loss information. Virtually all the information on aircraft losses exists because men who had served in British special forces wrote personal accounts of their wartime experiences. The destruction of Axis aircraft was a priority mission for them, so it should not be surprising that they kept good records of their successes. In contrast, destruction of aircraft was not a major objective for those ground units that captured airfields in the course of theater offensives. Histories of these latter operations discuss the major objectives, not the incidental destruction of aircraft. Thus, the number of incidents of ground attacks and aircraft losses on airfields in World War II could easily be double the numbers reported here. Recall that the Germans alone lost over 23,000 aircraft between May 1940 and June 1944.[30] It is easy to understand how chroniclers might have overlooked a few hundred aircraft lost to ground action over the course of five years of fighting on three continents.

The Vietnam data have a somewhat different problem in that they accurately capture aircraft losses but fail to count many attacks. Fox's listing of 475 attacks includes only attacks against the ten USAF main operating bases in the Republic of Vietnam. Other sources identify an additional 18 attacks against allied bases in Vietnam, Thailand, and Laos.[31] Attacks against facilities of the U.S. Army and other forces certainly number in the tens, if not the hundreds.

The next three chapters look at the two conflicts where all but 19 of the attacks occurred, beginning with the German attack on Crete in May 1941.

---

[30]Williamson Murray, *Strategy for Defeat: The Luftwaffe 1933–1945*, Maxwell AFB, Ala.: Air University Press, 1983, Table LXII, p. 304.

[31]Royal Laotian Air Force bases were also attacked by communist forces. See Events 10, 20, and 30 in Table B.3 of Appendix B.

Chapter Three

THE GERMAN AIRBORNE ASSAULT ON CRETE

This chapter discusses the successful German airborne assault of Commonwealth airfields on Crete in 1942. It is based on official histories and other scholarship on the conflict in the Mediterranean. It is the least quantitative of the three case studies because of the sources used, because it analyzes only one attack, and because of its purpose.

OVERVIEW

The primary strategic purpose of the attack on Crete was to prevent the Royal Air Force from launching long-range strikes against Axis forces in the Balkans. It had a secondary strategic purpose of extending the range of *Luftwaffe* aircraft in operations against the Royal Navy. The immediate tactical objective of the assault was to seize the airfields as airheads.

The loss of Crete convinced the British government that a dedicated air-base-defense force was needed and, in 1942, led to the creation of the RAF Regiment. U.S. Army Air Force leaders followed suit the same year and created a similar dedicated air-base-defense force,[1]

[1]The RAF Regiment remains a dedicated air-base-defense force, with organic air defense and field squadrons. The U.S. Army Air Force disbanded many of the dedicated air-base-defense units during WW II and the remainder at the end of the conflict. Since then it has relied on Security Police to provide air base defense in addition to their other duties. During the Vietnam War, the USAF experimented with the RAF Regiment model, creating the Safe Side Squadrons. The Safe Side units were essentially light infantry battalions assigned the sole mission of air base defense. They were

the Air Base Security Battalions. Thus, the attack on Crete is a seminal event for the RAF Regiment and a touchstone for USAF Security Police as well.

This experience is relevant for two additional reasons. First, there is some confusion in air-base-defense circles about what really happened on Crete. This chapter seeks to dispel several myths about the operation, and it offers lessons learned that are substantiated by the excellent historical materials available. Second, the British experience on Crete illustrates some problems associated with joint-service operations in defense of airfields. The following themes are developed:

- British leaders had advance knowledge of the details of the planned attack on Crete.
- New Zealand *ground* commanders on Crete failed to devote sufficient resources to the defense of Maleme Airfield.
- German airborne forces suffered terrible losses and were on the verge of defeat when New Zealand forces withdrew from Maleme.
- The popular image that RAF personnel were responsible for the loss of Maleme is incorrect.

BACKGROUND

In November 1940, British troops were dispatched to Crete to guard Suda Bay, thereby enabling the Greek government to deploy the Cretan 5th Division, the only Greek army unit on Crete, to the mainland, where it helped repel the Italian invasion. British forces on Crete did little to develop the island's defenses in the six months between November 1940 and May 1941 ("the calm before the storm"), despite Churchill's exhortation that Crete be turned into "another Scapa" (the heavily fortified Royal Navy Base in the Orkney Islands).

---

not used effectively in Vietnam; the program was disbanded in 1971. See Fox, 1979, pp. 110–113, for more on the Safe Side program.

In spring 1941, German leaders became concerned that the RAF might use Crete as a forward base to launch air strikes against Axis forces in the Balkans, representing a threat to the right flank of German forces as they advanced east in the upcoming invasion of the Soviet Union. In response to this threat, and driven by a desire to use his airborne forces in a large operation, German General Kurt Student developed a plan for an airborne assault on Crete. Student recognized that the three airfields on Crete were the key to a successful assault; his plan called for most of his forces to capture the airfields so that reinforcements and supplies could be flown in. Student understood that his lightly armed and outnumbered paratroopers would not last long without substantial reinforcements. The plan also called for some reinforcing units to arrive by sea.

To help bolster Greek defenses against the expected German attack, a Commonwealth expeditionary force of 30,000 British, Australian, and New Zealand troops was sent to Greece in early March. One month later, Germany invaded Yugoslavia and Greece. Greek and Commonwealth forces fought a delaying action until late April, when they were evacuated to Crete. By early May, Commonwealth forces on Crete totaled over 30,000 personnel and were supported by 10,000 Greek troops. None of these forces were particularly well equipped. Commonwealth forces had left much of their heavy equipment behind during the evacuation from Greece, and Greek forces were lightly equipped with antiquated weapons.

OPERATION MERCURY: MYTH AND REALITY

According to myth, the Germans achieved both strategic and tactical surprise. The story goes as follows. Commonwealth forces defending the airfields at Maleme, Retimo, and Heraklion on Crete were initially pinned down by intensive air attack. When the air attacks ended on the morning of May 20 and the defenders "came up for air they found themselves looking into the muzzles of tommy-guns"[2] in

[2]This quote is actually a somewhat idealized description of German airborne operations against Norway, Denmark, and Holland in May 1940, but it has been misused many times in the air-base-defense literature to describe *Operation Mercury.* The quote is originally from C. G. Grey, *The Luftwaffe,* London: Faber and Faber, 1944, p. 176. J. F. C. Fuller, 1948, used the Grey quote in a discussion of German airborne operations in his history of WW II. Numerous USAF authors from Fox to Bell (see the

the hands of the numerically superior German paratroopers. Royal Air Force personnel at Maleme were particularly ill-prepared to defend the airfield and were responsible for the loss of Maleme and, therefore, Crete. A typical view is that "British airfield personnel at Maleme, Crete, offered little resistance and quickly succumbed to a German attack."[3]

The historical literature is consistent in its refutation of this myth.[4]

First, Commonwealth forces were not surprised. Ultra, the British code-breaking team at Bletchley Park, England, had intercepted and decoded Hitler's April 25 message authorizing *Operation Mercury.* By April 29, the British War Office had received the details of the plan, which were communicated to Major General Bernard Freyberg,[5] the new Commander of Commonwealth Forces, Crete, on May 1. Furthermore, Freyberg received detailed updates from Ultra every few days before the battle and at least one message as late as May 21. Freyberg's information was so precise that when German gliders and parachutists appeared overhead, Freyberg observed that "they're dead on time."[6]

---

following footnote) then quoted Fuller, with the Grey quote embedded and unacknowledged, to describe the German assault on Maleme.

[3]Lieutenant Colonel John Ballard and Captain Jon Wheeler, "Air Base Vulnerability: The Human Element," *Air Force Journal of Logistics,* Summer 1989, p. 3. Similar assessments are found in Major General George Ellis, "More Hands for Base Defense," *Air Force Magazine,* December 1988, p. 69; Brigadier General Raymond Bell, "To Protect an Air Base," *Airpower Journal,* Fall 1989, p. 5; and in other Air Force staff and student papers.

[4]The most thorough and authoritative source is D. M. Davin, *Crete: Official History of New Zealand in the Second World War 1939–45,* Wellington, New Zealand: War History Branch, Department of Internal Affairs, 1953. See also Winston S. Churchill, *The Second World War, Vol. III: The Grand Alliance,* Boston: Houghton Mifflin, 1985 ed., pp. 238–269; Denis Richards, *The Fight at Odds, Volume I, Royal Air Force 1939–1945,* London: Her Majesty's Stationery Office, 1953, pp. 324–336; Antony Beevor, *Crete: The Battle and the Resistance,* Boulder, Colo.: Westview Press, 1994; George E. Blau, *The German Campaigns in the Balkans (Spring 1941),* Washington, D.C.: U.S. Army Center for Military History, 1953, pp. 119–147; and John Keegan, *The Second World War,* New York: Viking, 1989, pp. 160–172.

[5]Freyberg's previous command was the New Zealand Division. He was a lifelong infantryman and had received the Victoria Cross for his heroism in World War I.

[6]Ronald Lewin, *Ultra Goes to War: The First Account of World War II's Greatest Secret Based on Official Documents,* New York: McGraw-Hill, 1978, p. 157.

Second, the first wave of German forces landed near, but not on, the airfields. Many landed on Commonwealth ground force positions and were killed either as they descended or shortly after they landed. Two thousand died the first day, and an additional 1,986 were killed between May 21 and June 2. Three hundred and forty German aircraft also were lost, including 151 Junkers transports. British soldiers found it so easy to kill German paratroopers as they descended or struggled out of their harnesses that they used bird-shooting terminology to describe their successes.[7] German parachute harnesses and jump techniques contributed to this ease: The paratroopers were suspended by a single point, dangling like spiders and landing on their hands and knees. This landing method may have been effective in the Low Countries but produced many injuries on Crete's more rugged terrain. More important, German paratroopers were lightly armed, carrying only grenades and machine pistols; their rifles and machine guns were dropped in separate equipment bundles, often far from the troops who needed them.

Another point to keep in mind is that none of the Commonwealth airfields on Crete was functioning at the time of the assault. No RAF squadrons were permanently based in Crete until April 1941. When Group Captain G. Beamish arrived on April 17 to take command of RAF forces on Crete, construction was ongoing and there were no spares or repair facilities at any of the airfields. The small fields at Heraklion and Maleme could handle only fighter aircraft; Retimo was just a landing strip. As Greece was evacuated, surviving squadrons flew to Crete. In mid-to-late April, 14 Blenheim bombers, 14 Hurricanes, and 6 Gladiators arrived. Later that month, an additional 9 Blenheims arrived from Egypt. By mid-May—after the Blenheims were flown to Egypt—Crete's air force numbered 24 fighter aircraft.[8] Against this force, the Germans had arrayed 120 Do-17, 40 He-111, and 80 Ju-88 bombers; 150 Ju-87b Stuka dive bombers; and 90 Me-110 and 90 Me-109 fighters.[9] Little thought had been put into developing the island's defenses, and as late as May 5 the Chiefs of Staff were still debating whether and how to defend Crete. For example,

---

[7]Beevor, 1994, pp. 139, 230.

[8]Davin, 1953, p. 21; Richards, 1953, pp. 324, 326.

[9]Beevor, 1994, p. 347.

> Air Marshal Portal was emphatic that it would be dangerous to maintain an active air defence over the island at the expense of the Western Desert and elsewhere. The soundest course was to rely on AA [anti-aircraft artillery], dispersion and concealment, and at the same time to maintain a ground organisation which would permit aircraft to fly in from Egypt if seaborne attack was attempted.[10]

The British official history of the RAF in World War II sums up the dilemma faced by the RAF leadership:

> [A]t the most the three airfields could not have taken more than five squadrons of Hurricanes; and at the beginning of May there were not five Hurricane squadrons intact in the whole of the Middle East Command. Even if two or three squadrons could have been spared from their other tasks—which was virtually impossible with Malta under constant attack and Rommel on the borders of Egypt—they would still have been impotent against the overwhelming force of the enemy. To send further squadrons to Crete in the face of such odds was thus simply, in [Air Marshal] Tedder's view, to invite greater losses—losses which, coming on top of those incurred in Greece, might mean nothing less than the sacrifice of Egypt. The air commander accordingly resolved to maintain, if possible, a dozen Hurricanes in Crete, so that the enemy should not have matters all his own way; but he declined to expose more than this very limited number to the certainty of eventual destruction on the ground. His policy received the full support of the authorities at home.[11]

The RAF did take a number of prudent defensive steps. The incomplete airfield at Kastelli Pediados and open ground at Heraklion and Retimo were blocked by trenches and barrels of dirt. Revetments[12] were built to protect fighters, and each airfield was assigned a few howitzers, anti-aircraft guns, and light tanks.[13]

German air attacks against Suda Bay, ground forces, and the airfields increased in intensity in May. Every day the small RAF contingent

---

[10]Davin, 1953, p. 35.

[11]Richards, 1953, pp. 326–327.

[12]*Revetments* are walled enclosures built to protect aircraft from blast and fragmentation effects of nearby explosions. They are not covered and provide no protection from direct hits.

[13]Richards, 1953, p. 327.

battled with swarms of Me-109s, achieving many kills but steadily losing aircraft in the process. By May 19, only five operational fighters remained on the island. Rather than have the fighters destroyed, General Freyberg ordered them flown to Alexandria. Thus, the day before the German airborne assault, the airfields were no longer operational. From a doctrinal point of view, the airfields had ceased to be resources to be protected and should have been seen as key terrain to be denied to the enemy. It is surprising, then, that an all-out effort was not made to render the airfields unusable. Davin's analysis is instructive:

> According to Group Captain Beamish the intention was that the RAF should return in greater numbers and at a later stage. And although no document is available in which this is unequivocally stated, it seems clear that the view of the Chiefs of Staff was ultimately responsible. The result was that although every soldier near Maleme could see a case for destroying that airfield, it was obstructed but not destroyed. And, as events were to confirm, not to destroy the airfields was to make them more difficult to defend.[14]

The final element in the myth is the notion that RAF ground personnel were responsible for the defense of Maleme. In actuality, the 5th Brigade of the New Zealand Division was assigned the task of defending Maleme Airfield.[15] Indeed, with the evacuation of the last fighters on May 19, the RAF ground crews no longer had any reason to be on Crete and were not on the airfield on the day of the assault. (They were in the RAF camp just south of the airfield.) If anyone deserves blame for the loss of Maleme, it is the New Zealand 5th Brigade, not the RAF. Churchill's often-quoted and seemingly harsh words in his 1941 memo[16] should not be construed as evidence that RAF personnel at Maleme were responsible for its loss. Indeed, in Churchill's own history of WW II, he discusses the battle for Crete in detail; nowhere does he suggest that the RAF was responsible for the loss of Maleme.

---

[14]Davin, 1953, p. 51.

[15]Davin, 1953, p. 28.

[16]See Appendix A for the full text of Churchill's memo.

THE BATTLE FOR CRETE

The New Zealand Division (7,700 personnel) was deployed in the Maleme–Canea–Suda Bay area. Although its 5th Brigade was responsible for the defense of Maleme Airfield, the primary orientation of the division was to defeat seaborne forces coming into Canea or Suda Bay. The Australian 19th Brigade (6,500 personnel) was deployed at Georgioupolis and at Retimo Airfield. Most of the British 14th Brigade (15,000 personnel) was deployed around Heraklion Airfield, with one battalion assigned to Tymbaki on the southern side of the island.

German forces included the 7th Parachute Division and the 6th Mountain Division. The force was supposed to total almost 23,000, but heroic and costly attacks by the Royal Navy prevented 7,000 seaborne German troops from arriving on Crete.[17] In total, the attacking force of 16,000 faced 40,000 defenders.

On the morning of May 20, regiments from the German 7th Parachute Division assaulted Maleme and Canea, and an additional battalion was inserted by glider on the Akrotiri Peninsula (see Figure 3.1). Late in the afternoon of the 20th, two battalions parachuted

RANDMR553-3.1

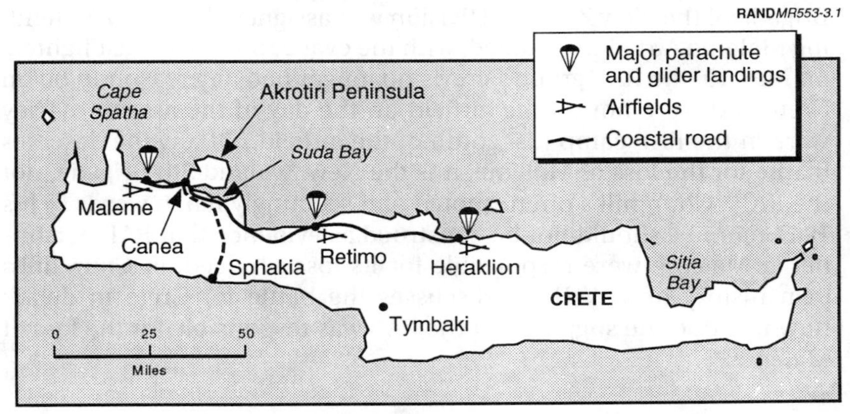

Figure 3.1—German Landings on Crete, May 20, 1941

---

[17]See Correlli Barnett, *Engage the Enemy More Closely: The Royal Navy in the Second World War,* New York: W. W. Norton and Company, 1991, pp. 352–364; and Beevor, 1994, Chapter 13.

into the Retimo area. At the same time, one regiment and one battalion attacked Heraklion. British commanders at Heraklion and Australian commanders at Retimo recognized that the early hours of the battle would be decisive. They committed their reserves early and dealt a terrible blow to the German forces.

In contrast, the New Zealand defense of Maleme was hampered by their obsession with the seaborne threat to Suda Bay. The New Zealand Division was strung out between Maleme and Suda and devoted only one brigade (the 5th) to the defense of Maleme. Furthermore, the 5th Brigade commander assigned only one of his three battalions (the 22nd) to the defense of the airfield, not nearly enough forces to defend 5 square kilometers of very rough and uneven terrain. The other two battalions (the 21st and the 23rd) were deployed to the southeast of the airfield. The 22nd Battalion was deployed with C Company on the airfield proper, HQ Company (organized and fighting as a rifle company) to the east, A Company on Hill 107 (the dominant terrain feature immediately to the south of the airfield), D Company to the west of Hill 107, facing the Tavronitis riverbed, and B Company on the east side of the hill. The 22nd Battalion also possessed two Matilda tanks located between Hill 107 and the airfield. The main weakness of this deployment was assigning only one company to cover Maleme Airfield. One platoon with only 22 men and no machine guns was responsible for the entire western side of the airfield, which was over 1 kilometer long.[18]

After several hours of intense air bombardment, German gliders and paratroopers began to land in the Maleme area around 8 a.m. on May 20 (see Figure 3.2). The Storm Regiment was assigned the mission of seizing Maleme. It was composed of 1 battalion of glider troops and 3 battalions of paratroopers. The gliders landed at the mouth of the Tavronitis River, west of the airfield; II Battalion landed well to the west, away from any defenders; IV Battalion landed just west of the Tavronitis River; and III Battalion landed to the southeast of Maleme.

Initially, the defense of Maleme went quite well. The German III Battalion landed on top of the New Zealand 21st and 23rd Battalion

[18]Davin, 1953, pp. 96–114.

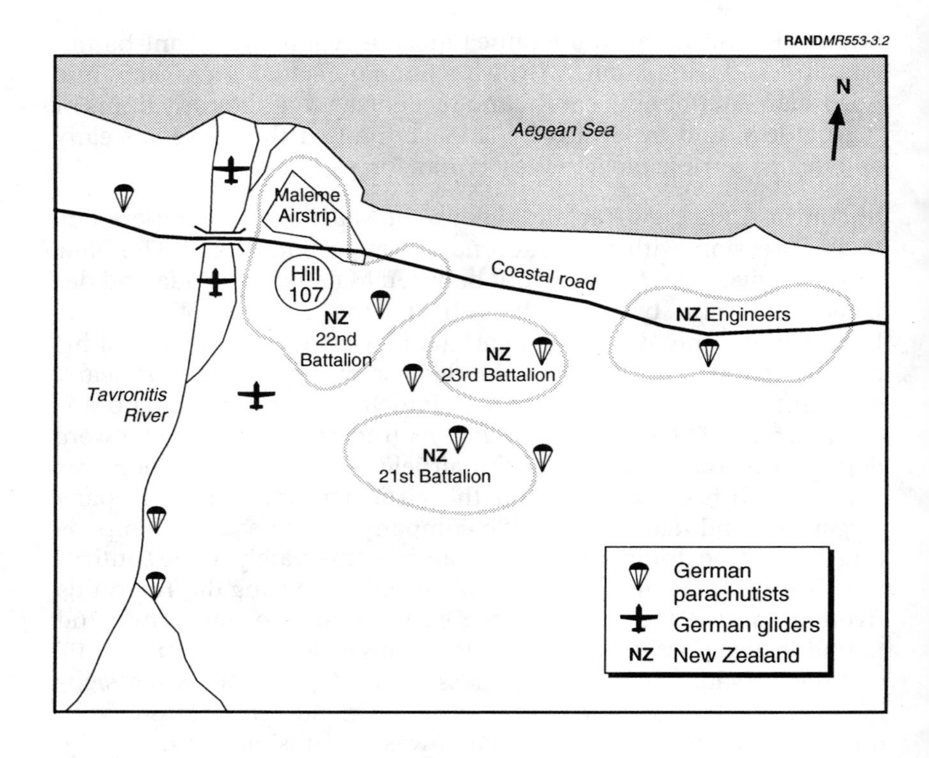

Figure 3.2—German Landings and Commonwealth Defenses at Maleme

positions southeast of the airfield and was virtually annihilated. Two-thirds of the German battalion's soldiers and all of its officers were killed.[19] To the west, Company D, 22nd Battalion, in positions on the west side of Hill 107, was able to fire effectively on gliders landing in the riverbed, but others landed in gulleys that hid them from view. With no radios and limited fire support, the defenders were not able to engage the out-of-sight forces. Captain T. Campbell, commanding D Company, recognized the importance of destroying the glider force before it had a chance to maneuver out of the riverbed. He requested that the 4-inch coastal defense guns on Hill 107 fire on the glider landing zones near the Tavronitis River

[19]Davin, 1953, p. 96.

bridge. Remarkably, his request was refused because those guns were "sited for targets at sea."[20] This bizarre incident—in which real targets were ignored in favor of potential targets—demonstrates both the myopia of Freyberg's defensive concept and the rigidity of the New Zealand command structure.

By 10 a.m., the German forces to the west had regrouped and were probing the western edge of the airfield. Colonel Leslie Andrew, the 22nd Battalion commander, refused a request from Company C's commander for armored support. Meanwhile, other Germans had crossed the Tavronitis bridge (which Company D's commander had requested artillery to fire on) and attacked the RAF camp located between C and D Companies. Although some of the RAF personnel were armed and had been given last-minute infantry training, they were not integrated into 22nd Battalion defensive plans nor were they under Colonel Andrew's control. By early afternoon, the RAF camp was overrun. 22nd Battalion's forward companies lacked radios, and, as the intensity of the fighting increased, runners had increasing difficulty reaching Colonel Andrew in battalion headquarters on Hill 107. By 11 a.m., Andrew believed that Companies C and D had been overrun. Andrew sent up flares, the signal for the 23rd Battalion to counterattack, but no help came. A runner got through to brigade headquarters—located 6 kilometers away—with a request for assistance, but no help came. Finally, Andrew spoke with his brigade commander at 5 p.m., requesting immediate reinforcements. Brigadier James Hargest, the brigade commander, incorrectly claimed that the 23rd Battalion was tied down by German forces and could not help. At this point, Andrew sent his two tanks and some infantry forward, but, rather than go to the airfield, they advanced to the west. One tank retreated after discovering that it had the wrong ammunition and that its turret would not traverse. The other advanced into the riverbed, where it became stuck on a large rock. At 6 p.m., Andrew contacted Hargest again, reporting that he would have to withdraw unless reinforced. Hargest said, "If you must, you must," but did promise to send two companies. When the reinforcements failed to arrive by nightfall, Andrew decided to withdraw. Beevor observes:

---

[20]Davin, 1953, p. 100.

> It is easy to sympathize with Colonel Andrew's state of mind, but harder to understand why he did not leave his command post, blind on the rear slope of the hill, and attempt to study the scene through binoculars. . . . [If he had done so] he would have seen that Captain Johnson and his men in C Company were still resisting strongly on the airfield, as was Campbell's D Company above the Tavronitis. They had suffered considerable casualties, but having inflicted far greater losses on the enemy, their thoughts were not on withdrawal.[21]

Andrew withdrew battalion headquarters during the night, but C and D Companies did not find out about the withdrawal until early morning. Their commanders were shocked to learn of this decision. Now isolated from the rest of their battalion, they had no choice but to withdraw also. By sunrise on May 21, no Commonwealth troops were left on Maleme. At 5 p.m. that evening, the first German transport aircraft landed on Maleme with reinforcements. The Germans flew reinforcements into Maleme over the next two days, with a transport unloading every minute at the peak of the airlift. The commander of the British coastal artillery battery on St. John's Hill asked for permission to traverse his 6-inch guns and engage the German troops on the airfield. However, like D Company of the 22nd Battalion, he was denied permission because those guns were to be reserved for anti-shipping missions only.[22] With no means of resupply and no air cover, Commonwealth forces had little prospect of reclaiming the lost airfields. Realizing this, Major General Freyberg ordered a withdrawal to the southern coast and the evacuation by sea of his command, leaving 5,000 troops and officers behind.

STRATEGIC EFFECT OF THE ATTACK ON CRETE

The German capture of Crete had two interesting outcomes.

First, it did achieve the German objective of denying Crete to the RAF, which is probably the less important outcome of the attack. Even if Crete had been held, it is unlikely that the RAF, at least in 1941 and 1942, could have conducted long-range bombing of any

---

[21]Beevor, 1994, pp. 124–125.

[22]Beevor, 1994, pp. 162, 149–154.

real consequence. It is true that Churchill placed a high priority on Mediterranean operations, but it seems unlikely that he would have diverted Bomber Command's scarce resources to Crete. In North Africa, the RAF was hard-pressed to support the campaign in the Western Desert and had no aircraft to spare for attacks against Balkan targets. In contrast, the *Luftwaffe* was able to make good use of Crete as a fighter and bomber base, mounting costly attacks on British shipping.

Second, perhaps the real strategic effect of the attack was a perverse one: The terrible losses suffered by German airborne forces on Crete convinced Hitler that large-scale parachute assaults were impractical, and he refused to authorize an assault on Malta. As described in Chapter Four, Hitler's failure to capture Malta and deny the RAF this crucial base was a major factor in the Axis defeat in North Africa.

CONCLUSIONS

Without German air superiority, *Operation Mercury* never would have taken place. If the RAF had been able to mount sustained attacks on German airfields in Greece or had sufficient fighters or antiaircraft artillery to counter German airpower, it is unlikely that *Mercury* would have gone forward. Yet despite the advantage that the *Luftwaffe* gave the attacking forces, Crete could have been held. If Freyberg had understood that the airfields, not the ports, were the center of gravity in the defense of Crete and had allocated his forces accordingly, the German assault would never have gained the foothold at Maleme. Maleme deserved protection by the entire 5th Brigade, not by a lone, unsupported battalion. Even this initial oversight could have been remedied if Freyberg, Major General Edward Puttick (commander of the New Zealand Division), or Brigadier Hargest had acted decisively and committed reserves early. The German forces were exhausted, and low on ammunition, water, and food. They undertook no offensive action against Maleme on the night of May 20 and could have been easily dislodged from their toehold on the airfield if a counterattack had been launched that night.

Churchill's June 14, 1941, memo to the Chiefs of Staff raises serious questions about Freyberg's leadership:

> I am far from reassured about the tactical conduct of the defence by General Freyberg, although full allowance must be made for the many deficiencies noted above. There appears to have been no counter-attack of any kind in the Western sector until more than 36 hours after the airborne descents had begun. There was no attempt to form a mobile reserve of the best troops, be it only a couple of battalions. There was no attempt to obstruct the Maleme aerodrome, although General Freyberg knew he would have no Air in the battle. The whole conception seems to have been of static defence of positions, instead of the rapid extirpations at all costs of the airborne landing parties.[23]

It is interesting to speculate how the fight for Maleme would have gone if the RAF Regiment had existed and had been deployed in field squadron or wing strength to Maleme. As noted above, Maleme was lost because of bad decisions, not because of a lack of forces. An RAF Regiment squadron assigned to Maleme would have had the advantage of being committed to the airfield and not subject to an army chain of command with other priorities. Additionally, the squadron would have been more likely than army units to appreciate the need to deny the field to the Germans and might well have blocked the landing strip, dug trenches, and placed explosives to render the field unusable. Additionally, they could be expected to have established heavy machine guns at both ends of the landing strip so that they could fire on any aircraft attempting to land. On the other hand, given his preoccupation with the seaborne threat, Freyberg probably would have used RAF defenders at Maleme as an excuse to move the 5th Brigade further east to protect the beaches at Canea and the port at Suda Bay.

No amount of heroic fighting by the troops could overcome the combination of initial maldeployments, lack of fire support and ready reserves, and unimaginative, out-of-touch commanders. Above all else, Crete was lost because of poor leadership. To attribute its loss to RAF ground crews may make good fiction, but it is a fundamental misreading of the historical record.

---

[23] *Churchill Memo to General Ismay for the Chiefs of Staff Committee*, June 14, 1941, quoted in Beevor, 1994, p. 229.

Within RAF and USAF circles, the Crete experience is cited as proof of what can happen when air forces are unprepared to defend their own bases. There certainly is truth in that observation, but confusion about the historical details has led many to focus on RAF shortcomings and to miss the equally important point that ground commanders are likely to have very different priorities from airmen.

The RAF experience on Crete vividly demonstrates the challenge of joint operations in defense of air bases, a problem that was not unique to the Commonwealth. As Chapter Four discusses, the *Luftwaffe* and *Afrika Korps* had equal difficulty coordinating defenses in North Africa.

Chapter Four

BRITISH SPECIAL OPERATIONS IN NORTH AFRICA AND THE MEDITERRANEAN

This chapter draws on personal and academic accounts of British special operations in North Africa to present an integrated assessment of the effectiveness of their attacks on German air bases, adding quantitative evaluations where possible.

OVERVIEW

British special forces in North Africa were the first ground forces to systematically attack enemy aircraft at airfields. Their operations are explored in some depth here for three reasons. First, the British Long Range Desert Group (LRDG), Special Air Service (SAS), and Special Boat Squadron (SBS; also referred to as the Special Boat Service) were highly successful, destroying almost 400 aircraft in a two-year period. Second, these operations are extensively documented in personal and academic accounts. Thus, in contrast to Viet Cong and NVA operations during the Vietnam War, we can learn much about the attacker's perspective on strategy, training, equipment, and tactics for air base attack. Third, this experience demonstrates what competent desert warriors can accomplish with quite primitive equipment. The German experience should be sobering for those who think desert environments protect air bases from attack by special forces.

The story of ground attacks on Axis air bases in North Africa is, above all, a story about inserting special forces hundreds of miles behind enemy lines. Because insertion was such a daunting problem for air base attackers and is likely to be so in many future wars, this discussion first describes desert-exploration efforts that gave the British the expertise needed to move motorized forces across hundreds of miles of trackless desert.[1] It then briefly reviews the formation of the various British special units that conducted the raids on Axis air bases. The core of the chapter describes and analyzes the 50 raids such units conducted between 1940 and 1943. Finally, some quantitative analysis of these operations is offered, looking for lessons learned for both attackers and defenders. The conclusion explores steps the Germans might have taken to disrupt British attacks. The following themes are developed:

- British innovation in adapting motor vehicles and navigation tools for cross-country travel in the desert and expertise in desert survival made these raids possible.
- Axis airfield defenses proved remarkably easy to penetrate.
- Small forces were able to cause great damage to Axis aircraft and made a major contribution to the Allied cause.
- Axis forces improved airfield defenses somewhat, but weaknesses in their rear-area defenses made them vulnerable to attacks during most of the campaign.

BACKGROUND

During World War I, British forces in Egypt began to experiment with motorized forces for desert reconnaissance and raids. The British faced a Turkish-supplied army of Bedouin tribesmen, the *Senussi*, who lived in oases running along the Libyan-Egyptian border. The camel-mounted *Senussi* would raid British and Egyptian outposts, then disappear into the desert. British forces lacked the mobility to respond to such raids; neither their cavalry nor infantry could operate in the vast desert surrounding the *Senussi* oases. Initially, the British sought to copy the *Senussi* with a Camel Corps of their own, which gave British forces mobility equal to the *Senussi*, but at the price of firepower, because a camel could not carry much more than man, rifle, food, and water.

---

[1]The next several pages draw on John W. Gordon's excellent history of desert exploration and warfare in North Africa: *The Other Desert War: British Special Forces in North Africa, 1940–1943*, New York: Greenwood Press, 1987, especially pp. 1–31.

A solution to this problem was found in 1914 Rolls Royce armored cars. These cars, which could carry four men and a mounted machine gun, were formed into Light Armored and Motor Machine-Gun Batteries. By 1916, these forces were augmented with Model T Fords carrying Lewis .303 machine guns and formed into the Light Car Patrols. The patrols were able to successfully cross the desert and began to attack the *Senussi* in their oasis homes. Denied the sanctuary of these oases, the *Senussi* were forced deeper into Libya and put on the defensive. Within a few months of these operations, the *Senussi* were no longer a significant threat.

In 1923, Italian forces developed their own motorized desert unit—the Auto Sahara Company—and used it with great success in their efforts to put all of Libya under Italian control. The British, meanwhile, had disbanded their specialized desert warfare units. Several British officers and one civilian living in Egypt did, however, have a great interest in desert exploration. The efforts of this small group did much to advance the use of motorized transports for desert travel.

Captain Ralph Bagnold, a signals officer stationed in Egypt, pioneered the use of motorized vehicles across the worst of the desert terrain. Bagnold and other junior officers combined their personal funds to purchase Model T Fords, equipment, and supplies. The vehicles were modified with radiator condensers to capture water when the radiators boiled over and sun compasses for navigation;[2] windshields, hoods, and other nonessential parts were removed to lighten the cars.

In 1926, Bagnold's group completed a 1,000-mile trip into the Sinai Desert, the first of many ambitious desert explorations. In 1929, they attempted a crossing of the dunes of the Egyptian Sand Sea, a feat considered impossible at the time. When the expedition reached the eastern edge of the Sand Sea, they looked up at a 400-foot-high "glaring wall of yellow sand." Bagnold drove one of the Model Ts straight to the top. Mechanical problems and insufficient gasoline reserves forced the expedition to turn back, but they had demon-

---

[2]Bagnold describes the challenge of desert navigation and his invention of the sun compass in his *Libyan Sands: Travel in a Dead World*, London: Hodder & Stoughton, 1935, pp. 84–86, 154–155.

strated that motorized vehicles could cross the dunes. The group returned in 1930 and made the first crossing of the 200-mile-wide Egyptian Sand Sea in a remarkable 3,000-mile loop. William Kennedy Shaw, a civilian official with the Sudan Colonial Service and a crack desert navigator, joined them for this trip. Shaw used *dead reckoning* (taking the direction from his sun compass and distance from his car's odometer) during the day and double-checked their position at night with star sightings from a theodolite.[3] These amateur explorers—Bagnold, Shaw, Patrick Clayton, and Guy Prendergast—would later lead the British desert patrols of World War II.

THE LONG RANGE PATROLS

In spring 1940, General Archibald Wavell, Commander of Commonwealth Forces in the Middle East, faced an Italian force in Libya reportedly five times larger than his own. Although Italian ground forces and air bases were concentrated on the narrow coastal strip and adjacent Libyan Plateau, there were reports that the Italians were setting up bases in the south to conduct long-range bombing of British facilities in Egypt and the Sudan. Also, Captain Bagnold, whom General Wavell had reassigned to Egypt, thought that the Italians might use their Auto Sahara unit to seize the Siwa Oasis in western Egypt and conduct raids against the Nile Valley.[4]

The British, however, lacked good intelligence on Italian activities, particularly those at the key Italian base at Kufra in southeastern Libya. Bagnold proposed creating a motorized reconnaissance-and-raiding unit whose first mission would be to infiltrate Libya via the Egyptian Sand Sea and observe Italian traffic on the Palificata, the track leading to Kufra. An analysis of traffic on this, the sole route to Kufra, would provide Wavell crucial intelligence on the units operat-

---

[3]Gordon, 1987, p. 28.

[4]During his 1932 expedition to Jebel-al-Uwaynat, 700 miles south of Cairo, Bagnold and crew had a chance encounter with an Italian Auto Sahara company. The two groups shared a meal and discussed desert operations. An Italian major by the name of Lorrenzini kidded Bagnold that if Britain and Italy ever went to war, the Auto Saharans could launch raids from Uwaynat and strike targets at will in the Nile River Valley, including the Aswan dam and the Sudanese railhead at Wadi Halfa. See Gordon, 1987, pp. 29–31.

ing there. Additionally, Bagnold proposed that such a unit could "exploit the advantages of a secret route deep into the heart of Libya,"[5] conducting raids and reconnaissance missions against Axis installations throughout Libya.

Wavell, who had known T. E. Lawrence (Lawrence of Arabia) years before and was a bit of an unconventional warrior himself, was attracted to this relatively inexpensive way of countering the numerically superior Italian forces and authorized the formation of a special force to be called the Long Range Patrols (LRP). The LRP would consist of a headquarters and three patrols, designated R, W, and T. Each patrol would have 30 men and 12 vehicles. The new unit subsequently acquired from a host of sources its critical equipment: 3/4-ton, 1-1/2–ton, and 5-ton trucks; theodolites, sun compasses, sunglasses; .303 caliber Browning machine guns, World War I–vintage Lewis guns, and some antitank guns; and radios.[6]

By September 1940, the LRP had established a small base camp at Big Cairn on the western edge of the Egyptian Sand Sea on the Libyan-Egyptian border. At the same time, the Italians launched an offensive into Egypt. The LRP was ordered to begin combat operations immediately. On the LRP's first raid, W patrol discovered several Italian auxiliary airstrips north of Kufra and destroyed fuel dumps and pumping facilities at each airstrip.

In October and November 1940, all three patrols were sent out. The W and R patrols were to attack the Italian airstrip and outpost at Ain Zwaya. W was spotted and attacked by Italian Ghibli bombers, but R patrol discovered and destroyed an unguarded Savoia S.79 Italian bomber, the first Axis aircraft in North Africa to be destroyed by Commonwealth special forces. W patrol also destroyed 8,000 gallons of aviation fuel and three tons of aircraft munitions stored nearby. T patrol went north, crossing the Egyptian Sand Sea and planting mines within 70 miles of Agedabia. T patrol also ambushed convoys

---

[5]Gordon, 1987, p. 43.

[6]Gordon, 1987, pp. 43–51.

and raided outposts along the way, completing a 2,100-mile circuit in 15 days.[7]

The strategic purpose of these operations was to distract General Graziani, the Italian commander, and force him to divert forces to protect his airfields and outposts. Graziani, an old desert hand who had created the Auto Sahara company a generation earlier, understood the threat and feared it. As frantic messages from his outposts continued to come in, it appeared that British forces were hitting targets simultaneously throughout Libya. Consequently, General Graziani pulled ground forces and aircraft away from his offensive for rear-area security. The Auto Saharans did successfully ambush T patrol south of Kufra, capturing Captain Clayton. Surprisingly, they never did engage in their own raids against British rear areas, nor did they seek to track down and destroy the LRP in its desert base camps.

General Wavell was so pleased with the successes of the LRP that he expanded its operation from three to five patrols and renamed it the Long Range Desert Group, seeking to increase the pressure on Graziani and buy time to ready British forces for a major counteroffensive. In December, British forces struck hard at the Italians, driving them back to the Libyan border. By February 1941, British forces had advanced to the Gulf of Sirte. At the same time, the first elements of Rommel's *Afrika Korps* began arriving in Tripoli, a development that would change the course of the campaign. (See Figure 4.1.)

On January 11, 1941, the LRDG conducted its most ambitious operation to date, raiding the airfield and fort at Murzuk in western Libya. Murzuk, the Italian district headquarters for the Fezzan province, was over 700 miles from the nearest Commonwealth units. The LRDG patrol set fire to a hangar and other facilities, destroying three Italian light bombers.[8]

---

[7]R. L. Kay, *Long Range Desert Group in Libya: 1940–41*, Wellington, New Zealand: War History Branch, Department of Internal Affairs, 1949, p. 6; W. B. Kennedy Shaw, *Long Range Desert Group*, London: Greenhill, 1945 (1989 ed.), p. 49; Gordon, 1987, pp. 57–58.

[8]Shaw, 1989 ed., pp. 53–67.

RANDMR553-4.1

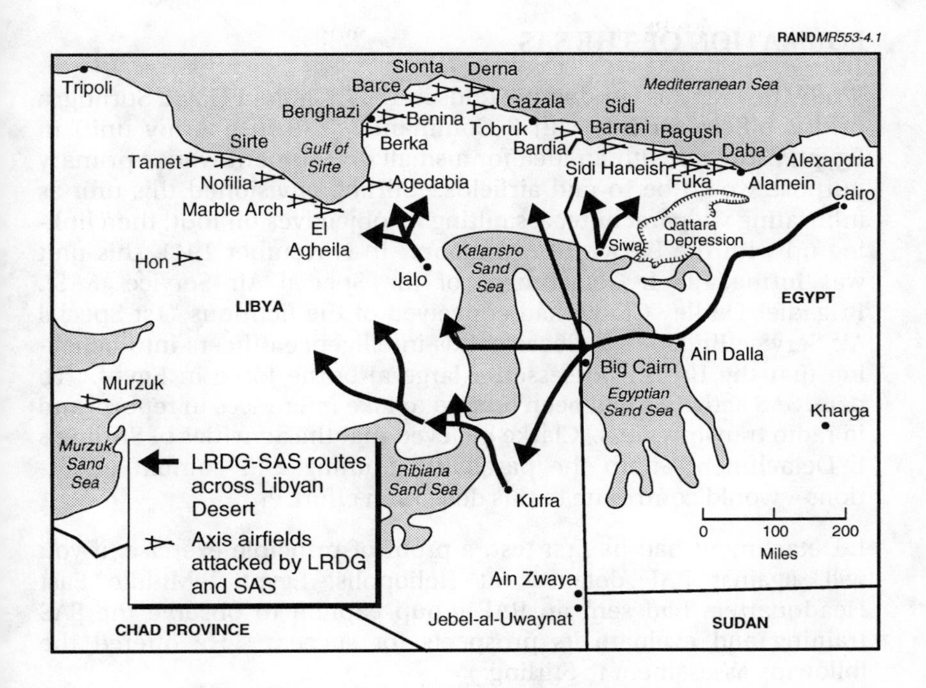

Figure 4.1—Map of Libyan Desert

In March, Rommel's *Afrika Korps* struck British forces at El Agheila. The LRDG collected intelligence and harassed German forces as British forces fell back to Egypt. After a summer lull caused by logistics difficulties and garrison duty guarding the Kufra oasis,[9] the LRDG shifted operations to the north, where it transported agents to the coastal strip and conducted its own surveillance missions. By September, the new commander of the LRDG had concluded that 30-man, 12-truck patrols were too large and visible for the heavily patrolled Libyan Plateau, and reorganized them into 10 12-man, 5-truck patrols.[10]

[9]Kufra was captured in March 1941 by the LRDG and Free French. The Free French garrison was being pulled back to Chad, and the LRDG was the only force available to guard this valuable outpost.

[10]Gordon, 1987, p. 76.

FORMATION OF THE SAS

While the LRP was conducting raids in 1941, Captain David Stirling, a British officer serving with 8 Commando (a British Army unit) in Egypt, came up with an idea for a small detachment whose primary purpose would be to raid airfields. Stirling envisioned this unit as infiltrating via parachute, assaulting its objectives on foot, then linking up with the LRDG for extraction. In November 1941, this unit was formed as L Detachment of the Special Air Service (SAS). Brigadier Dudley Clarke had conceived of the fictitious "1st Special Air Service Brigade" to deceive Axis intelligence officers into believing that the British possessed a large airborne force in Egypt. To date, SAS activities had been limited to fake references in reports and in radio transmissions. Clarke believed that the activities of Stirling's L Detachment—both the parachute training and combat operations—would contribute to this deception effort.[11]

L Detachment had its first test, a proof-of-principle exercise, if you will, against RAF defenses at Heliopolis, Egypt. Middle East Headquarters had sent an RAF group captain to observe the SAS training and evaluate its prospects for success. He offered the following assessment to Stirling:

> A very innocent view you take, Captain Stirling; a presumption in fact. I can't speak for the enemy, though I'm sure that their defences will be every bit as good as our own, but I assure you that you wouldn't find ours an easy target.[12]

Stirling could not resist this challenge and proposed a test of his unit against the RAF defenses at Heliopolis air base near Cairo, Egypt. The SAS, carrying full operational loads—including food, water, and rocks representing satchel charges, infiltrated on foot the 90 miles from its base at Kabrit, 100 miles northeast of Cairo on the shore of the Great Bitter Lake. The force moved only at night, hiding under camouflage netting during the day to avoid detection by RAF recon-

---

[11]Alan Hoe, *David Stirling: The Authorized Biography of the Creator of the SAS*, London: Warner Books, 1992, p. 70.

[12]Hoe, 1992, p. 88.

naissance aircraft. Alan Hoe describes its arrival at Heliopolis four nights after departing Kabrit:

> A smooth operation then followed as they snipped through the outer perimeter fence and, unobserved, made their way on to the airfield and amongst the parked aircraft. Mayne's group placed between forty and fifty labels [surrogates for bombs] on the aircraft and before they retreated they moved around the target as much as possible just to prove to themselves that it could be done.[13]

L Detachment's first mission was in support of the Commonwealth *Crusader* offensive. The operational plan called for the SAS to parachute into Libya, attack five airfields, then link up with an LRDG motorized patrol for the return to Egypt. On the night of November 16, 1941, five aircraft departed Egypt for drop zones near five airfields in the vicinity of Gazala, Libya. The aircraft became lost in a storm and dropped the raiders far from their objectives. One aircraft tried to raise RAF air controllers by radio. German controllers, using perfect RAF procedures and proper English, responded and tricked the pilot into landing at Gazala, where everyone on board was captured. Two planeloads were dropped far to the south, probably in the Egyptian Sand Sea, and were never found. In total, 38 members of the 62-man force were killed or captured. None reached their objectives.[14] Fortunately for Stirling and the other survivors, the LRDG was waiting for them at the assigned rendezvous and returned them safely to Egypt.

The November debacle caused Stirling to rethink his operational concept and to conclude that airborne insertion was too unpredictable. Stirling had been impressed by the competence and reliability of the LRDG and proposed that the LRDG and SAS team up on future raids. Major Guy Prendergast, the LRDG commander, agreed. For 10 months, from December 1941 until September 1942, the LRDG and SAS jointly planned and conducted raids throughout Libya, operating initially out of a joint forward base at the Jalo oasis,

---

[13]Hoe, 1992, pp. 89–90.

[14]John Strawson, *A History of the S.A.S. Regiment*, London: Secker and Warburg, 1984, p. 252; Gordon, 1987, pp. 81–84.

150 miles southeast of the Gulf of Sirte and within 200 miles of most Axis airfields. (See Figure 4.1.)

DECEMBER 1941 RAIDS

The first combined LRDG-SAS operation targeted Axis airfields at Sirte, Tamet, El Agheila, and Agedabia. Stirling took one man (Sergeant Jimmy Brough) with him to Sirte. Once at the airfield, they conducted a pre-raid survey of the airfield, penetrating the defenses and counting many Italian light bombers. They did not plant charges on the aircraft, because the plan called for simultaneous attacks the next night on Sirte and nearby Tamet. Sergeant Brough describes what happened next:

> We tripped over a couple of sleeping men, off-duty guards, I think. They started firing—then other guards started and then anti-aircraft guns opened up. None of the bullets were coming our way though. Captain Stirling reckoned they must have thought they were being attacked from the sea and said it was a good thing they seemed more scared than we were. We made a place to lay up back on the ridge and when daylight came we found we were in the middle of a bunch of Arab women grubbing around with mattocks [a digging tool]. We were very still for about three hours and they went away. That afternoon we noticed that the aircraft (we'd counted about thirty Eyetie Capronis) kept taking off in pairs and nothing seemed to be landing. By late afternoon, the airfield was empty. It looked as though they'd been flown away for safety after we'd disturbed the sentries.[15]

Mayne's group penetrated the Tamet perimeter without detection, attacked a guard house or crew quarters, then destroyed 24 aircraft and a fuel dump with satchel charges. Captain Jock Lewes and his 9-man team found no aircraft at El Agheila. (The Italians had recently

---

[15]Hoe, 1992, pp. 110–111. Hoe claims that the aircraft were evacuated to nearby Tamet, where they were destroyed by Mayne's team. Perhaps, but none of the accounts of the attack on Tamet mentions a dramatic inflow of aircraft. Furthermore, Mayne's team destroyed only 24 aircraft. There should have been more aircraft at Tamet if the 30 aircraft from Sirte had deployed there. Another possibility is that Sirte was being used as a staging base and the aircraft were just passing through. This is David Lloyd Owen's view in *Providence Was Their Guide: A Personal Account of the Long Range Desert Group, 1940–45,* London: Harrap, 1980, p. 70.

moved them to Agedabia.) Captain Bill Fraser's 4-man patrol slipped through Agedabia's extensive barbed-wire barriers without detection by the many guard posts, but then had to lie still for over an hour, waiting for an opportunity to get onto the airstrip. Once there, they planted charges, withdrawing in the confusion following the first detonations. It is interesting that Fraser's group found that the detonations helped their escape. In several later attacks, premature detonation of charges led to detection of the SAS forces, which then had to fight their way off the field. Fraser's team destroyed 37 aircraft, including Italian CR-42 fighters, Stukas, and Me-109s. These first four raids destroyed 61 aircraft and 25 trucks without a single SAS casualty.[16]

Upon returning to Jalo, Stirling decided to launch another series of raids immediately. He sent Fraser's team to Marble Arch, Lewes' to Nofilia, and Mayne's back to Tamet. Stirling returned to Sirte, which he found well guarded by German armor, and was unable to get onto the field. Fraser's team found Marble Arch empty. Lewes' team penetrated Nofilia. Sergeant Bob Lilley, a member of the team, describes the operation:

> We reached Nofilia before dawn (26 December) and found a place to hide up where we could watch the aerodrome. There were not many aircraft in the field and the few that were there were very widely dispersed, but we noted the positions of them. As soon as it was dark we moved on to the landing ground and put a bomb on the first plane. We had just put one on the second when the first one went off—we were only using half-hour time pencils then. After that the airfield became alive with troops and we came very near to getting caught as we beat a retreat. It was a disappointment to all of us that we had only destroyed two planes.[17]

MARCH 1942 RAIDS

On March 8, 1942, the SAS launched its next series of airfield raids. Stirling and two men went to Benina several times, discovering that it

---

[16]Hoe, 1992, pp. 110–112; Strawson, 1984, pp. 45, 253; Shaw, 1989 ed., p. 124; Gordon, 1987, p. 91; Warner, 1971, p. 46.

[17]Strawson, 1984, p. 45.

was a major repair facility with no operational aircraft assigned. They did, however, manage to destroy two torpedo dumps. (The torpedoes were used against Allied convoys, particularly those to Malta.) Mayne and three men took on Berka's satellite field, destroying 15 aircraft and an equal number of torpedoes. One team member was captured. The team assigned to Slonta was unable to penetrate the perimeter.[18] The team assigned to Berka No. 1 could not find the airfield. Finally, Fraser's patrol attacked Barce, destroying 1 aircraft and some trucks. In late March, Stirling returned to Benina and, using the knowledge gained from his earlier reconnaissance and raid, destroyed 5 aircraft in hangars. Most of the aircraft on the field were decoys.[19]

JUNE 1942 RAIDS

Both Axis and Commonwealth forces in North Africa depended on resupply by ship. Most supplies for Commonwealth forces came up the Red Sea, although some urgent convoys, such as the 1941 *Tiger* convoy bringing desperately needed tanks, took the more hazardous but shorter Mediterranean route. The vast majority of Axis supplies came from Italy to the Libyan port of Tripoli; port limitations and vulnerability to RAF attacks curtailed the use of Benghazi and other ports to the east. As a result, air, land, and sea operations were tied closely. For example, British airfields on the Island of Malta served as way stations for almost half of all RAF aircraft destined for the Middle East. Malta played a vital offensive role, as well: Its aircraft, ships, and submarines severely disrupted Axis efforts to supply their forces in the Western Desert, sinking, between July and December 1941, 581,000 tons of ships carrying armor, trucks, fuel, ammunition, and food destined for the *Afrika Korps* and *Luftwaffe*, and helping to stall Rommel's offensive at the Egyptian border.[20] In turn, Malta was de-

---

[18]The accounts of this raid do not say specifically what perimeter defenses (e.g., guards, mines, lights, fences) turned them away.

[19]Johnny Cooper, *One of the Originals: The Story of a Founder Member of the SAS*, London: Pan Books, 1991, p. 42; Strawson, 1984, pp. 55, 254; Hoe, 1992, pp. 142–143.

[20]Rommel faced four major problems in his supply lines: fuel shortages that prevented necessary convoys from sailing; the vulnerability of convoys at sea; Tripoli's limited port capacity; and the extreme distance from the port to the front lines. Each problem dominated his logistics nightmare at some point during the war. For exam-

pendent on RAF fighters based in North Africa to protect its resupply convoys from attack by Axis aircraft and ships. When the 8th Army did well and airfields in northeast Libya were in British hands, Malta prospered. When these airfields were in German hands, as they were in the late winter and spring of 1942, Malta suffered greatly.[21] During the worst of this onslaught on Malta's ports and airfields, Axis aircraft flew about 6,000 sorties and dropped almost 7,000 tons of bombs.[22]

By June, Malta was in desperate need of resupply. Two convoys were planned for the second week of June, one from Alexandria and one from the United Kingdom via Gibraltar. Middle East Command took several steps to help the convoy, including intensively bombing Axis airfields and asking for the assistance of the SAS. In response, Stirling devised a plan to attack the key Axis airfields on Crete (see Figure 4.2) and in North Africa while the convoys were at sea.

Infiltrating by submarine and raft, and working with partisans on the island, three SBS and one SAS team used the techniques developed in North Africa to attack four airfields on Crete. The SBS team assigned to attack Maleme Airfield was turned back by impressive defenses that included many machine-gun posts, dogs, and searchlights. Another SBS team found Tymbaki abandoned. A third SBS team attacked Kastelli Pediados, using Lewes bombs[23] to destroy 8 aircraft and almost 200 tons of aviation fuel and other supplies. At Heraklion, the team climbed up a hill and counted 60 Junkers 88

---

ple, when the lines stabilized at El Alamein, Rommel's forces were 1,300 miles from the port of Tripoli. *Afrika Korps* trucks could transport only 300 of the 1,500 tons needed every day to keep the forces at the front supplied. Furthermore, Tripoli could handle only 45,000 tons per month compared with the 70,000 tons the total Axis force required. See Martin Van Creveld, *Supplying War: Logistics from Wallenstein to Patton*, Cambridge, England: Cambridge University Press, 1977, pp. 181–201.

[21]Excellent discussions of the battle for Malta can be found in Cajus Bekker, *The Luftwaffe War Diaries*, Garden City, N.Y.: Doubleday, 1968, pp. 233–253; Matthew Cooper, *The German Air Force: 1933–1945: Anatomy of Failure*, London: Jane's, 1981, pp. 203–212; John Strawson, *The Battle for North Africa*, New York: Charles Scribner's Sons, 1969, pp. 108–109; and Correlli Barnett, *Engage the Enemy More Closely: The Royal Navy in the Second World War*, New York: W. W. Norton and Company, 1991, pp. 492–510.

[22]Kreis, 1988, pp. 117, 123, 131.

[23]The *Lewes bomb* was an incendiary device designed by SAS Captain Jack Lewes. It became the SAS weapon of choice.

RANDMR553-4.2

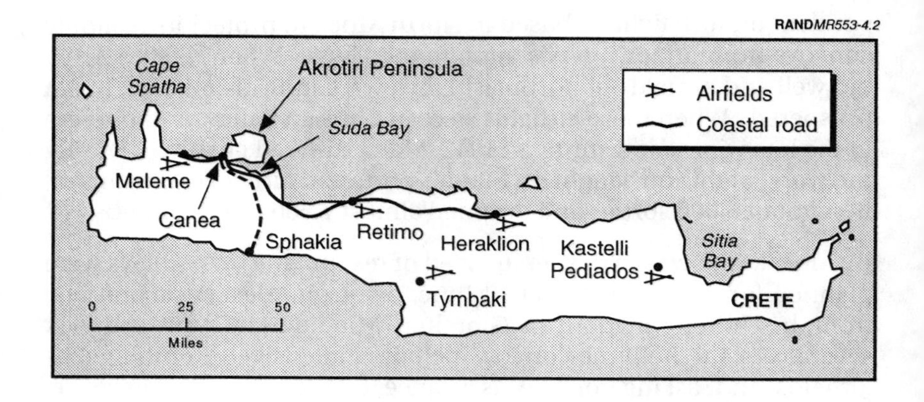

Figure 4.2—German Airfields on Crete, June 1942

(Ju-88) bombers dispersed in revetments around the field. Later that same night, a sentry challenged the team and opened fire. The team withdrew and decided to try again the next night. The next night, the team cut a hole in the perimeter fence, then hid in a small shed as a German patrol passed outside of the wire. The Germans discovered the hole and began to climb through to search for the intruders. In a remarkable and lucky coincidence, a single RAF bomber made an attack run on the airfield at that moment. As the bombs exploded, the SAS slipped away and onto the southern side of the airfield, where they placed Lewes bombs on 21 bombers. As they crossed the airstrip to the northern side, they expected problems with the roving patrols but found they could be avoided quite easily. On the northern side of the airfield, they discovered that most of the aircraft were non-operational. Rather than waste their bombs, they placed them on aircraft engines being repaired, and on some trucks and fuel supplies. The team watched their charges detonate and were surprised that few of the aircraft caught fire. Later, they learned that the aircraft tanks were drained (and presumably vented) as an anti-sabotage measure. All but one member of the team were captured or killed the next day.[24]

---

[24]Hoe, 1992, pp. 175–177; Strawson, 1984, pp. 107–108, 255; Beevor, 1994, pp. 261–262; Warner, 1971, p. 57; Lodwick, 1990, pp. 34–37; and Ladd, 1983, p. 30.

In North Africa proper, Stirling sent teams against five airfields. Stirling and two men attacked Benina, carrying 60 Lewes bombs between the three of them. They slipped onto the airfield, then waited for an RAF diversionary attack on Benghazi. The RAF raid had three purposes in addition to distracting the guard force. First, the guard force usually stood-to during an air raid, and the resulting activity allowed the SAS to identify most of the guard posts. Second, the raid also provided illumination[25] that helped teams see the aircraft and hangars. Third, teams typically would take compass bearings on the targets during a daylight reconnaissance or during air raids. Once darkness had returned, they would follow their compasses to the targets. Without such aids, they found that one could walk right by an aircraft. Once they had set the fuzes on their bombs, they planted two bombs on an aviation fuel dump; dodged a sentry; then, leaving one man outside as a guard, went inside a hangar. The hangar interiors were neither lighted nor guarded. Johnny Cooper, one of the team members, describes the operation:

> As our eyes grew accustomed to the darkness we saw that the hangar was full of German aircraft. Motioning me to go to the right, David set off to the left, and we busily placed our bombs on the Stukas and Messerchmitts that were in there for repair. . . . We then continued to the second hangar and dealt with more aircraft while Reg discovered a mass of spare aero engines and highly technical looking equipment. All this accounted for our full stock of sixty bombs.
>
> While still behind the last hangar we heard our first bomb go off, to be followed in rapid succession by all the rest. Although all the time pencils had been activated at the same time, there was always a slight difference due to differing acid strengths. Almost deafened by the noise we struggled through a gap in the wire, crossed the main road and scrambled up an escarpment. About three hundred feet up . . . we sat down and watched the fantastic firework display. The effect was stupendous. The real highlight probably came from the Me110 in the first hangar. When this caught fire, the heat from the burning fuel caused the 20mm cannons to go off spontaneously. Brilliantly coloured tracer shot across the airfield, giving the appearance of an anti-aircraft gun in action. The petrol dump went

[25]SAS raids were planned for moonless nights, when possible.

> up and the hangars were ablaze. Complete pandemonium reigned among the Germans on the base. When the real anti-aircraft guns opened fire, it appeared that they thought they were in the middle of an air raid.[26]

At Berka's main airfield, the SAS destroyed 11 aircraft with satchel charges, then, following discovery by sentries, fought their way out in "a raging and highly mobile battle during which they accounted for a large number of guards and amazingly suffered no severe casualties."[27] At Berka's satellite field, the SAS destroyed 1 aircraft.

In total, the SAS raids on June 9–13 destroyed 61 aircraft.[28] This was an impressive accomplishment; the destruction of 21 Ju-88 bombers and 200 tons of fuel in the sanctuary of Crete was a painful blow for the *Luftwaffe*. Perhaps even more important was the destruction of some of the small fleet of Ju-52 transports at Benina. Rommel relied heavily on these aircraft for priority shipping and could not afford to lose a single one. These raids, therefore, were important victories for the SAS, but they do not appear to have helped the convoys reach Malta.

The *Harpoon* convoy from the United Kingdom was first attacked by Italian and German aircraft based in Sardinia and Sicily, two locations untouched by SAS raids. A total of 200 Axis aircraft and a large Italian naval force attacked the convoy on June 14 and 15, sinking four of the six resupply ships as well as several Royal Navy escort ships.[29] It is possible that North Africa–based aircraft were part of the attacks on the 15th (when the convoy was within their range). If so, the SAS raids on Benina and Berka would have limited the number of Axis aircraft available for attacks, but not significantly. All 20 aircraft destroyed at Benina were under repair when they were destroyed on June 13. It seems unlikely that they would have been available for operations on the 15th. With an operational availability rate of roughly 50 percent, only 6 of the 12 aircraft destroyed at

---

[26]Cooper, 1991, pp. 52–53.

[27]Hoe, 1992, p. 171.

[28]This is my count. Stirling claims 75 destroyed. See Strawson, 1984, p. 62.

[29]Fortunately for the garrison on Malta, the two surviving ships contained sufficient stores to last until the next resupply attempt in August.

Berka's airfields would typically have been flyable. The contribution of 6 more aircraft to an attacking force of 200 would have been negligible. The SAS also destroyed fuel at the Barce airfield, although the amount and effect on operations are unknown.

The *Vigorous* convoy from Alexandria put to sea on June 13. On the 14th, it was attacked by Axis aircraft based in Crete and North Africa. Air and sea attacks forced the convoy to turn back and flee to Alexandria. The SAS raids on Crete may have saved the convoy from additional damage; they cannot, however, be credited with aiding Malta in any direct fashion in June 1942.[30]

JULY RAIDS

The LRDG and SAS launched a series of raids on Axis airfields throughout the month of July.[31] At Bagush (Figure 4.1), half of their bombs failed to detonate. On the spot, Stirling and Mayne decided to use their jeep-mounted machine guns to shoot aircraft. Stirling recollects:

> There we were with guns aboard which were designed for the RAF to shoot down aircraft in the air—why couldn't we do just that from the ground and keep the bombs in reserve? Paddy was all for it and we decided simply to drive on to the field and shoot the beggars up. It was amazingly easy—it was a total surprise to the Jerries. We used only the Blitz Buggy[32] and one jeep. We did a circuit of the

---

[30]Barnett, 1991, pp. 505–510.

[31]The British had one advantage in these and subsequent raids against German airfields in Egypt: They were all in areas where British airfields had been located prior to their capture by Rommel and the stabilization of front lines near Alamein. For example, in November 1941, RAF squadrons were based at Sidi Haneish (Landing Grounds 13, 101, and 102), Fuka (Landing Grounds 16, 17, and 103), Bagush (Landing Grounds 103, 115, and 116), and Sidi Barrani (Landing Ground 75). It is not clear that Axis aircraft operated out of all (or any) of the landing grounds, but it seems likely that they would have taken advantage of at least some of these fields. Surprisingly, accounts of the July SAS operations do not mention whether they did. Although Axis engineers and ground defense personnel probably made many changes to the airfields, the SAS must have benefited from operating in an area well known to British forces. See Kreis, 1988, pp. 146–147.

[32]The Blitz Buggy was a Ford V-8 convertible modified to look like a German staff car.

> perimeter and poured down as much lead as we could. We left the whole place littered with burning planes.[33]

This bold new technique destroyed 15 planes on top of the 22 already destroyed by satchel charge.[34] This success, the SAS acquisition of 30 jeeps, and Stirling's concerns that the Axis would act to counter his infantry penetration techniques led to the development of more refined jeep raid tactics.

On July 26, 1942, Stirling and a force of 50 British and French SAS in 18 jeeps used this technique in a raid on Sidi Haneish, destroying 40 aircraft. The attack destroyed many Ju-52 transports, aircraft in short supply and used by Rommel for high-priority resupply missions. Malcolm James, the unit medical officer, describes the attack:

> It looked as if they had caught the enemy completely red-handed, for as they drew near, they could see that the flare-paths were lit up. Some aircraft were being loaded whilst others were awaiting their turn to take off; and others still were circling low overhead watching for a clear runway. . . . The raiding party approached cautiously, and then, when they had drawn close, they accelerated and opened up with everything they had got. . . . At one moment the airfield had been a hive of German efficiency, with everything running to schedule—at the next, there had been utter chaos and confusion. . . . Within a few minutes there were fires and explosions everywhere. Aircraft were burning like matchwood. Each was riddled with incendiary bullets until it burst into flames.[35]

Using Stirling's new technique, the LRDG conducted its own jeep raid on Bagush in early August. This raid destroyed 15 ME-109s. It is worth noting that the Bagush defenders had not improved their defenses against these tactics in the month since Stirling's first ad hoc jeep raid on the field.

---

[33]Hoe, 1992, pp. 181–182.

[34]Cooper, 1991, pp. 58–62; Strawson, 1984, p. 64.

[35]Malcolm James, *Born of the Desert*, London: Greenhill Books, 1991, p. 160.

THE FINAL RAIDS

On September 12, two 4-man teams from the British Special Boat Squadron were inserted by submarine on the Island of Rhodes. Their mission was to destroy German and Italian bombers used to attack Royal Navy convoys. Local guides provided information about the airfield defenses, food, and water. At both Maritza and Calato airfields, the teams penetrated the perimeter and placed bombs on aircraft, munitions, and fuel. At Maritza, the team made an assessment the morning after the raid, reporting "many burnt-out aircraft."[36] The accounts do not give the number of aircraft destroyed, but the descriptions suggest that approximately 20 aircraft were destroyed.[37]

One day later, on September 13, 1942, the LRDG conducted the last air base raid in North Africa against Barce airfield.[38] This raid used the jeep assault technique, stopping to open the front gate on the way in. It destroyed 32 aircraft, and hangars and fuel trucks.[39]

In October 1942, General Bernard Montgomery launched his attack on German positions west of El Alamein. By late December, Commonwealth forces had pushed the *Afrika Korps* to the western edge of the Gulf of Sirte, capturing most of the air bases that the SAS and LRDG had attacked so successfully. The new front lines were extremely dense, with Axis airfields concentrated in the narrow coastal strip and surrounded by German forces. This density of enemy forces, an unfriendly native population, and extremely difficult terrain along the southern flank ended the remarkable string of air base raids.

In June 1943, three SBS teams returned to Crete. Tymbaki was found abandoned again, which is surprising, given the excellent intelli-

---

[36]Ladd, 1983, p. 34.

[37]Ladd, 1983, pp. 32–35; Lodwick, 1990, pp. 40–44.

[38]The LRDG did try to raid airfields at Hon and Sibha in January 1943 but had to abort the mission because heavy rains had made the vehicle approaches impassable. See Kay, 1950, p. 11.

[39]Shaw, 1989 ed., pp. 201–203; James, 1991, p. 278; Gordon, 1987, p. 129; and Peniakoff, 1950, p. 149.

gence the British had on German activities on Crete.[40] Another team planned to attack Heraklion, but their Cretan guide warned them that the airfield was rarely used, recommending instead that they blow up a fuel dump at Peza. The team assigned to Kastelli Pediados found that defenses had improved considerably since 1942: Three guards were now assigned to each Stuka (considered a high-value aircraft), one taking a turn on duty while the others slept in a tent next to the aircraft.[41] A diversionary attack created sufficient confusion that the other team members were able to plant their charges. They destroyed 3 Stukas, 1 Ju-88, and 1 observation plane.[42]

In July 1943, the SBS raided the German airfield at Ottana, Sardinia, destroying several aircraft.

ANALYSIS OF ATTACKS

How successful were British attacks on Axis airfields? They destroyed at least 367 aircraft, plus repair facilities, tons of ammunition, thousands of gallons of fuel, and many spare engines.[43] During June 1942, the SAS destroyed 8 percent of German aircraft based in North Africa.[44] Table 4.1 shows the breakdown of aircraft destroyed, by raid and location; most were destroyed between June and September 1942. (See Figure 4.3.)

A typical SAS-LRDG operation would involve a platoon-size force. Upon arriving near the objectives, the SAS element would separate into assault teams assigned to the targeted airfields. As Figure 4.4 shows, these teams were as small as 2 men and as large as 50 men, but 4–6-man teams were the most common. The SAS discovered through experience that 5-man teams were ideal for most targets;

---

[40]Beevor, 1994, p. 259.

[41]Lodwick, 1990, p. 62.

[42]Philip Warner, *The Special Air Service,* London: William Kimber, 1971, pp. 94–95; Beevor, 1994, p. 285; and Lodwick, 1990, p. 63.

[43]This is my count based on a review of the 53 attacks. Cooper (1991, p. 81) claims over 300 destroyed, and Hoe (1992, p. 226) reports over 400 destroyed. Ladd (1983, p. 29) credits Paddy Mayne with personally destroying 130 aircraft.

[44]My calculation. German air order of battle is from Kreis, 1988, p. 157.

Table 4.1

Axis Aircraft Destroyed by British Special Forces in North Africa and the Mediterranean, 1940–1943

Date	Location	Aircraft Destroyed
Oct 1940	Ain Zwaya	1
Jan 1941	Murzuk	3
Jan 1941	Kufra	1
Dec 1941	Tamet	24
Dec 1941	Agedabia	37
Dec 1941	Tamet	27
Dec 1941	Nofilia	2
Mar 1942	Berka	15
Mar 1942	Barce	1
Mar 1942	Benina	5
Jun 1942	Benina	20
Jun 1942	Berka No. 1	11
Jun 1942	Berka No. 2	1
Jun 1942	Kastelli Pediados	8
Jun 1942	Heraklion	21
Jul 1942	Bagush	37
Jul 1942	Fuka	14
Jul 1942	Fuka	22
Jul 1942	Sidi Haneish	40
Jul 1942	Unknown	15
Sept 1942	Barce	32
Sept 1942	Rhodes	20
Jun 1943	Kastelli Pediados	5
Jul 1943	Sardinia	5
TOTAL		367

they were small enough to be stealthy and large enough to provide mutual support.[45]

Stirling and his officers were quite deliberate in their attacks on aircraft:

[45]Information on team size is not available for every raid.

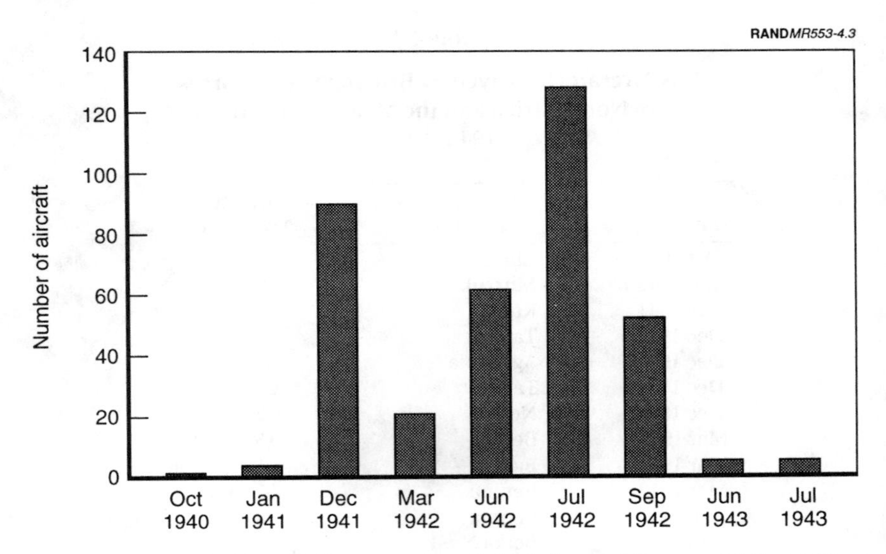

Figure 4.3—Axis Aircraft Destroyed by British Special Forces, 1940–1943

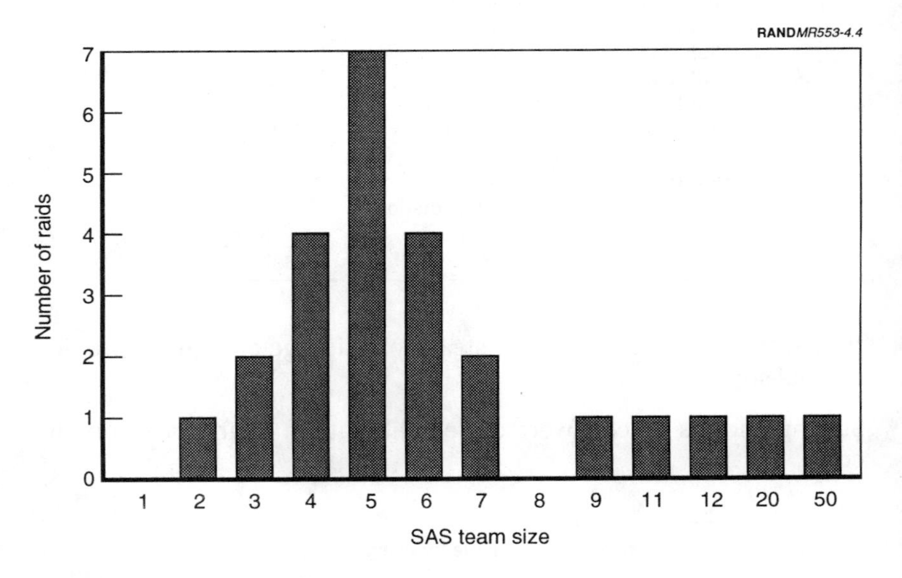

Figure 4.4—Variations in SAS Team Size

> The SAS developed a fine technique of limiting possible cannibalisation. If the best method to destroy an aircraft explosively on the ground was to place the bomb on the wing, how much more of a nuisance it would be if all bombs were placed on the port wings. No way then that the enemy can retrofit the "spare" wings to other aircraft.[46]

It is also interesting that they set the timers on their bombs at the beginning of the raid rather than individually at each plane. None of the sources explained why they did this. We can speculate on their reasoning. This approach had the advantage of producing near-simultaneous explosions, but at the risk of the team's being left holding armed explosives if they had to interrupt their activities. The bombs had a pin that was pulled at the last minute so that team members did not have to worry about the bombs going off in their rucksacks. Yet unanticipated delays in the operation could result in explosions before the team had left the airfield, rendering bombs in their packs dangerous and useless.[47]

LRDG/SAS successes were due in large part to their ability to use the desert as a sanctuary. Conversely, most Axis forces, lacking the skills and equipment necessary to travel far from the coastal strip, found the desert impenetrable. Axis aircraft were the only serious threat to the LRDG once it was south of the Libyan Plateau. Even then, the threat was greatest when LRDG patrols were on the move the morning after a raid; when stopped and camouflaged, the small patrols were exceedingly difficult to detect.

As Figure 4.5 shows, the majority of British raids were conducted on weekend nights. Although the day of the week was only one of many mission-planning considerations—strategic imperatives, coordination with other operations, and phase of the moon being among the most prominent considerations—it does appear that the SAS tried to launch raids on weekends when possible. Attack timing is not mentioned in any of the books on the SAS, but it seems reasonable that

---

[46]Hoe, 1992, p. 165.

[47]See Hoe, 1992, pp. 164–165, and Cooper, 1991, pp. 52–53, for descriptions of the Lewes bombs.

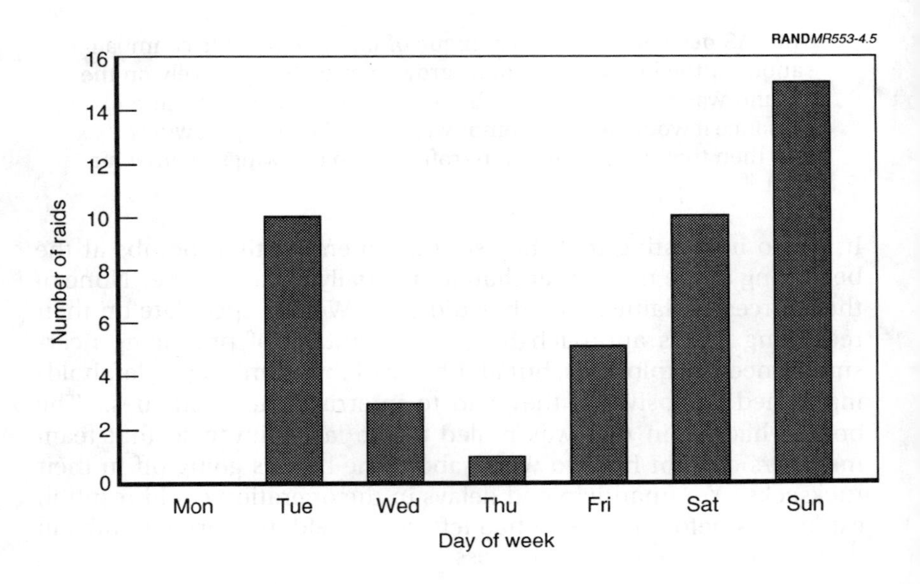

Figure 4.5—Timing of British Attacks on Axis Airfields

the SAS would have sought to exploit any tendency of Axis forces to relax on weekends.[48] Of the 44 raids for which we have specific dates, 30 occurred on the weekend as opposed to the 18 raids one would expect if the distribution were random. None of the raids happened on a Monday. Thirty of the 53 raids were successful—a 57 percent success rate. Raids that did not destroy aircraft, equipment, or stores were stopped by the problems shown in Figure 4.6.

After the SAS's great run of successes, Stirling began to think of new techniques for air base attack. As discussed earlier in this chapter, one major innovation was the jeep raid. Stirling observes:

> I was concerned that once the Germans fully caught on to our tactics it would be quite simple for them to make life very difficult for us. Certainly it would tie down manpower, but all they had to do

---

[48]Evidence from multiple conflicts suggests that, whenever the tactical circumstances allow, Western armies give passes, organize recreational sports activities, hold worship services, and generally rest and relax on weekends.

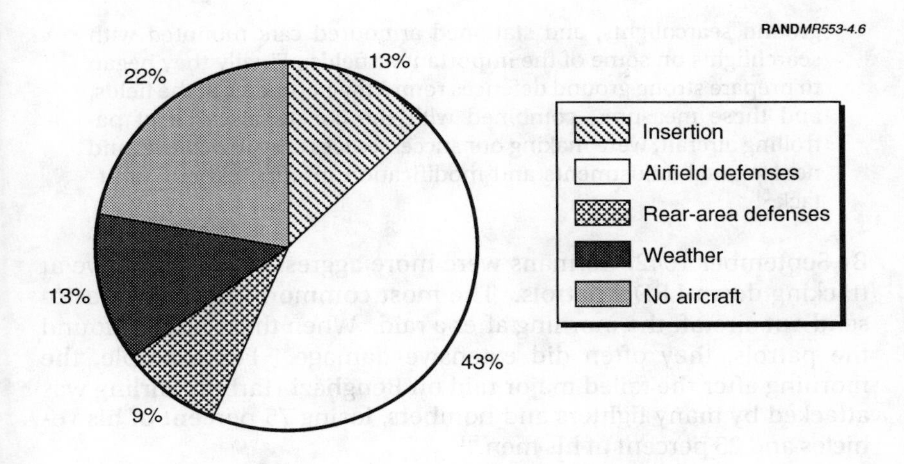

Figure 4.6—Causes of Unsuccessful Air Base Attacks

> was double up on their airfield sentries, keep them alert and we would have problems getting in. Taking it one step further, they knew we had to escape over the desert and that in some key areas the routes were limited; a well mounted and armed shut-off force would have severely limited us. If we could keep them guessing and use a variety of tactics in such a manner that they never knew what was coming where, and how it was being delivered—we could sow real confusion in the rear.[49]

Malcolm James, L Detachment's unit physician, suggests that the Germans were already making the changes Stirling feared:

> At first they thought to counter the destruction of aircraft by making a man sleep beneath the wing of each plane, but the subsequent loss of life had convinced them that this was not the correct solution: it had resulted merely in a loss of one man per plane, and presumably there were quite a few of the ground staff who suffered from uneasy sleep at nights. Then they decided to strengthen their guards; that made it more difficult for our men to approach the planes, but the raids were still successful. Later they fixed up

[49]Hoe, 1992, pp. 184–185.

> ground searchlights, and stationed armoured cars mounted with searchlights on some of the important airfields. Finally they began to prepare strong ground defences round the perimeter of the fields, and these measures, combined with the constant search of patrolling aircraft, were making our successes less easy to achieve, and necessitated adjustments and modification in our methods of attack.[50]

By September 1942, Germans were more aggressive and effective at tracking down LRDG patrols. The most common technique was to send out aircraft the morning after a raid. When these aircraft found the patrols, they often did extensive damage. For example, the morning after the failed major raid on Benghazi Harbor, Stirling was attacked by many fighters and bombers, losing 75 percent of his vehicles and 25 percent of his men.[51]

AXIS LESSONS

After the highly successful July 1942 raids that destroyed over 100 Axis aircraft, a German armored-car regiment made an unsuccessful attempt to find the SAS rendezvous point. The Germans also mined some avenues of approach. Occasionally, those mines damaged a vehicle and injured personnel.[52] Yet all those efforts were ad hoc, purely reactive, and lacked integration. What could Axis forces have done to improve their airfield defenses?

Lessons Learned—Organizational

Organizationally, it would have helped if the *Luftwaffe* had been subordinate to Rommel as the theater commander. Two parallel and often competing lines of command undermined joint air operations[53] and, although the evidence is sketchy here, appear to have

---

[50]James, 1991, p. 156.

[51]Hoe, 1992, p. 198.

[52]James, 1991, pp. 193, 229, 231.

[53]See Kreis, 1988, pp. 153–156, for a discussion of the effect of the *Luftwaffe–Afrika Korps* split on air support for ground operations.

hampered the coordination of rear-area defenses. Relations with the Italian allies were often strained, further undermining coordination. Indeed, the SAS was able to exploit this lack of integration by posing as Italians when among Germans and as Germans when among Italians.

In addition to the chain-of-command problem, there was an organizational division of labor for base defense. Interior defense of *Luftwaffe* bases was the responsibility of *Luftwaffe* base defense units, whereas general rear-area security belonged to the *Afrika Korps*—a logical way to organize base defenses, but one that requires careful coordination to avoid gaps in coverage, something the *Luftwaffe* and *Afrika Korps* lacked. An additional problem at Italian bases was the uniformly low quality of the base defense forces. *Luftwaffe* base defense units appear to have been considerably better.

Lessons Learned—Tactical

At the tactical level, more aggressive patrolling of the southern boundary of the coastal strip and Libyan Plateau would have hampered LRDG and SAS movement. In particular, night ambushes on likely avenues of approach (10–30 miles from the airfields) might have caught the raiders well before they reached the airfields. A well-laid ambush along such an approach—employing mines, heavy machine guns, and antitank guns—would have been detrimental indeed for the raiders in their unarmored vehicles. At the air bases, minefields, dogs, lights, and more sentries would certainly have made penetration more difficult. Again, night ambushes and listening posts along the wadis and other approaches would likely have been effective in detecting and stopping SAS teams infiltrating on foot.

Passive measures had proved valuable elsewhere and should have been employed in North Africa. For example, dispersing aircraft in revetments would have significantly increased the time the SAS forces were exposed and would have limited damage from a single explosion—an important point, considering that the SAS bombs on aircraft often ignited nearby fuel stores, weapons, and other aircraft.

Finally, emptying and venting aircraft fuel tanks—as the Germans did on Crete—would have minimized the damage from incendiaries.

STRATEGIC EFFECT OF THE ATTACKS

SAS operations caused a significant loss of aircraft and materiel and routinely disrupted Axis airfield operations. In a campaign characterized by a tenuous air balance and plagued by shortages, SAS destruction of 367 aircraft, fuel stores, munitions, and spare parts made a significant contribution to the British cause. Furthermore, SAS operations cost the Commonwealth relatively little in manpower and materiel. Conversely, SAS activities cost the Axis more than the aircraft and materiel destroyed, because they often caused aircraft and ground forces to be diverted from other missions to search for the raiders and manpower to be tied down guarding bases. Just as Axis commanders never fully appreciated the damage that Malta was doing to their cause, it appears that neither the *Luftwaffe* nor Rommel fully appreciated the damage caused by the SAS. In at least one letter home, Rommel expressed admiration for Stirling, describing him as "the very able and adaptable commander of the desert group which had caused us more damage than any other British unit of equal strength."[54] Rommel failed, however, to take any significant steps to stop these attacks.

CONCLUSIONS

The British experience in North Africa demonstrates what determined, competent, desert warriors can accomplish against a complacent air force. Modern air forces operating in desert environments should not assume that the desert environment or great distances from the front lines by themselves afford protection from small-unit attack. For example, during the 1990–1991 Gulf War, several important allied air bases were closer to Iraq and Yemen than most Axis bases were to British lines. The Iraqis proved to be a poorly motivated, uncreative, and often incompetent adversary; it would be a mistake to assume that all future foes will be equally weak. Future foes will not necessarily use SAS tactics; to the extent

[54]B. H. Liddell Hart, ed., *The Rommel Papers,* London: Collins, 1953, p. 393.

that they do, however, the countermeasures discussed above remain valid.

The next chapter leaps ahead 20 years, to the Vietnam War, in which small forces were also very successful against air bases.

USAF photo, courtesy ot the U.S. National Archives **RAND***MR553-P.1*

A survey of the damage at Tan Son Nhut, Vietnam, after the April 13, 1966, attack by mortars and recoilless rifles. A Vietnamese C-47 transport is in the center.

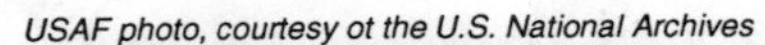

USAF photo, courtesy ot the U.S. National Archives **RAND***MR553-P.2*

Remains of an F-4C destroyed by rocket attack, July 15, 1967, Da Nang AFB, Vietnam.

USAF photo, courtesy ot the U.S. National Archives **RAND***MR553-P.3*

All that remains of revetments and the aircraft they sheltered after a rocket attack at Da Nang, Vietnam, in February 1967.

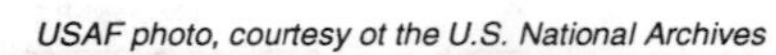

USAF photo, courtesy ot the U.S. National Archives **RAND***MR553-P.4*

Revetments constructed at Bien Hoa, Vietnam.

USAF photo, courtesy ot the U.S. National Archives **RAND***MR553-P.5*

USAF F-100 destroyed when a 107-mm rocket made a direct hit on its uncompleted shelter, February 23, 1969, Bien Hoa AFB, Vietnam.

USAF photo, courtesy ot the U.S. National Archives **RAND***MR553-P.6*

A concrete-reinforced shelter, such as this one with an F-4 at Phu Cat AFB, Vietnam, protected aircraft from damage during a rocket attack at Da Nang in March 1969.

Photo courtesy of UPI/Bettman **RAND** *MR553-P.7*

Charred metal is all that remains of an A-7 Corsair after terrorists bombed Muñiz Air National Guard Base, Puerto Rico, on the morning of January 12, 1981.

Photo by Frank Camacho, NYT Pictures **RAND** *MR553-P.8*

Terrorists destroyed these A-7 jets at Muñiz Airport, Puerto Rico, on January 12, 1981.

Chapter Five

AIR BASE ATTACKS IN VIETNAM AND THAILAND

This chapter discusses Viet Cong (VC) and North Vietnamese Army (NVA) attacks on USAF main operating bases (MOBs) in the Republic of Vietnam and Thailand during the 1964–1973 period. It draws on Air Force Project CHECO reports on air base attacks (including a recently declassified report on attacks against USAF bases in Thailand), the official USAF history,[1] and other official and unofficial reports. Beginning with an overview of attacks during the entire war, the chapter next discusses selected attacks against MOBs, the USAF concept of operations for base defense, and attacker tactics. Finally, it presents a comparative analysis that identifies the most at-risk bases and explores some of the factors that contributed to the problems of their defense.

OVERVIEW

The detailed records on base attacks in Vietnam provide an opportunity to do more quantitative analysis than was possible in the preceding two case studies. For this reason, and not to repeat historical narrative that already exists in the Project CHECO reports and Fox's book, this chapter is the most quantitative of the three case studies. It develops the following themes:

- USAF air base defenses at the MOBs were highly effective in detecting and stopping penetrating attacks.

[1] Roger Fox, *Air Base Defense in the Republic of Vietnam: 1961–1973*, Washington, D.C.: U.S. Air Force Office of History, 1979.

- The VC and/or NVA launched only 21 sapper attacks, and those attacks caused relatively little damage to USAF aircraft.
- Ninety-six percent of the attacks on MOBs used standoff weapons rather than attempting to penetrate defenses.
- Standoff attacks proved extremely difficult to stop.
- Additional ground and air patrols were needed to control the standoff footprints.[2]

BACKGROUND

More ground attacks on air bases were recorded in Vietnam than in any other conflict. VC and NVA forces attacked USAF main operating bases 475[3] times between 1964 and 1973. Those attacks destroyed 99 U.S. and Vietnamese aircraft and damaged another 1,170.[4] Additional attacks against other USAF, Army, Marine Corps, and Republic of Vietnam Air Force (RVNAF) facilities in Vietnam and against USAF bases in Thailand raised the total destroyed to 375, roughly 4 percent of all aircraft losses.[5] Although this is a relatively small percentage of the total losses, it is interesting that more U.S. Air

---

[2]The *standoff footprint* is the area around a base from which weapons can be fired onto aircraft and other targets. Its size varies with the type of weapon; typically, it extends 10 kilometers beyond the perimeter fence.

[3]Eighteen additional attacks against air bases were identified in the course of this research. Five of these were against MOBs in Thailand and are discussed in this chapter. The other 13 attacks are not discussed here; short descriptions can be found in Appendix B.

[4]Fox (1979) reports 100 destroyed and 1,203 damaged, but his database adds up to 99 destroyed and 1,170 damaged. For example, see p. 173, where he reports 36 RVNAF aircraft damaged in 1966, yet the individual entries for each attack show only 2 aircraft damaged.

[5]With a few exceptions, data for *damaged* aircraft at these other locations were not available. Also details for most of these other attacks in the Republic of Vietnam, including their locations, dates, and nature of attack, were not available. Such inconsistencies in the depth and breadth of data require that the discussion of air base attacks and associated damage differentiate between attacks on MOBs and those on other locations. Other sources of data on losses are USAF, *USAF Management Summary: Southeast Asia Review*, February 28, 1974, and Rene Francillon, *Vietnam: The War in the Air*, New York: Arch Cape Press, 1987.

Force fixed-wing aircraft were destroyed by ground action than were downed by MiGs (99 versus 62).[6]

As Figure 5.1 shows, ground attacks on air bases generally followed the trend of the entire war: As the war escalated from 1965 to 1968, so did the number of attacks; as the war slowed from 1971 to 1973, the number of attacks also dropped. The number of attacks increased dramatically from 17 in 1967 to a wartime peak of 121 in 1968. The next two years each saw over 100 air base attacks, but attacks in 1971 and 1972 dropped to roughly 50 per year.

Figure 5.2 shows the damage caused by these attacks following a similar progression: almost a tenfold increase in destruction from 1965 to 1966, a slight increase in 1967, followed by a threefold increase in 1968, with over 500 aircraft damaged or destroyed.

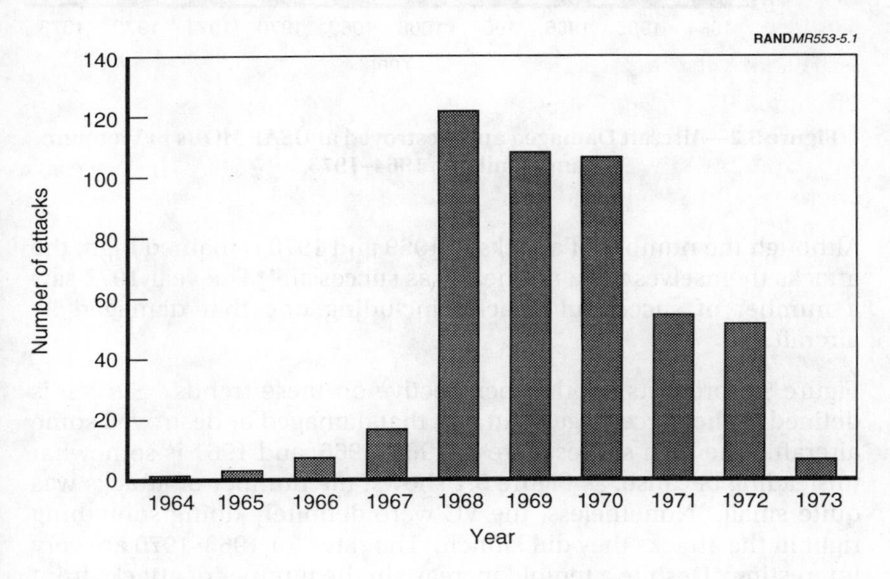

Figure 5.1—Attacks Against USAF Main Operating Bases (MOBs) in Vietnam and Thailand

---

[6]I am indebted to my colleague Chris Bowie for this observation. See Francillon, 1987, Table 2.

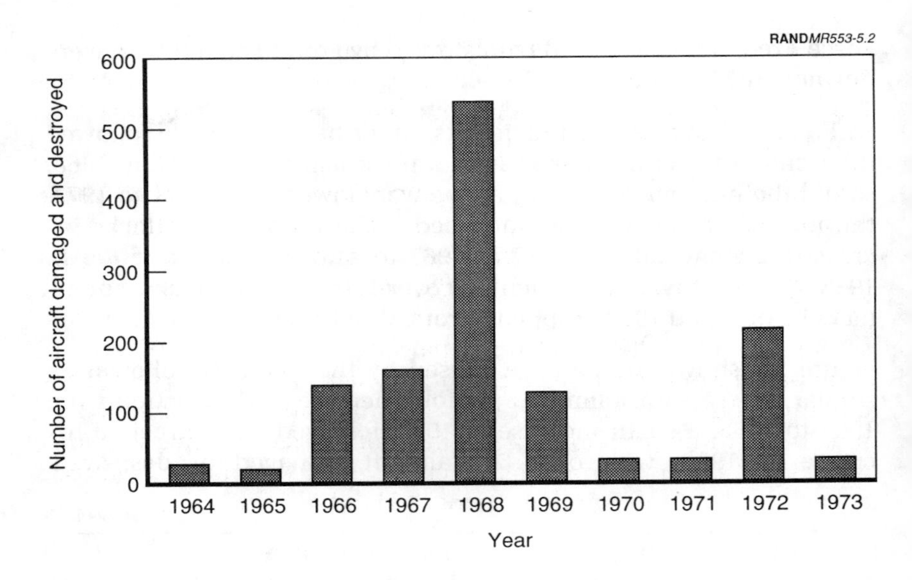

Figure 5.2—Aircraft Damaged and Destroyed at USAF MOBs in Vietnam and Thailand, 1964–1973

Although the number of attacks in 1969 and 1970 remained high, the attacks themselves were not nearly as successful. The year 1972 saw a number of successful attacks, including one that damaged 94 aircraft.

Figure 5.3 presents another perspective on these trends. *Success* is defined as the percentage of attacks that damaged or destroyed some aircraft. The high success rate in 1965, 1966, and 1967 is somewhat misleading because, as Figure 5.1 shows, the number of attacks was quite small. Nonetheless, the VC were definitely doing something right in the attacks they did launch. The rates for 1968–1970 are very interesting: Despite a tenfold increase in the number of attacks from 1967 to 1968, the success rate fell.

The relatively low success rates for 1968–1971 to some extent reflect the cumulative effect of defensive countermeasures instituted between 1965 and 1968. Increased surveillance by air made it more difficult to infiltrate large forces near the bases, and the use of gun-

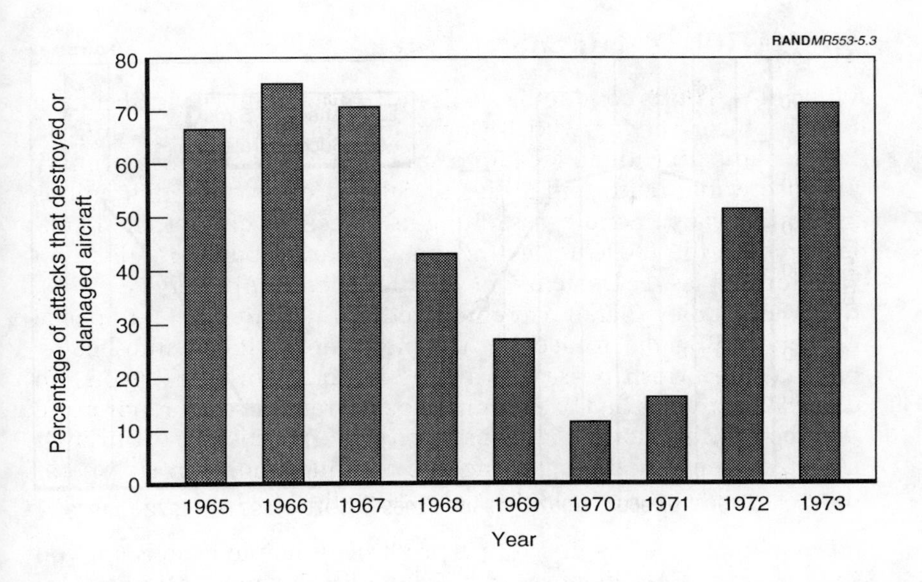

Figure 5.3—Attack Success Rate Against MOBs, 1965–1973

ships and counterbattery fire[7] rendered larger, more prolonged attacks quite dangerous for the attackers. It is also the case that the VC/NVA lacked the resources and trained personnel to successfully execute many large operations. *The combination of defensive countermeasures and the attackers' own resource limitations caused the average attack size to drop dramatically.*

Figure 5.4 presents for comparison the percentage of attacks during which more than 5 rounds (rds) were fired and the success rate. As one might expect, the success rate tracks quite nicely with the size of attacks. In 1970 and 1971—the least successful years for the attackers—only 24 and 20 percent of the respective attacks fired more than 5 rounds. It appears that the attackers learned from these experiences, because in the last two years of U.S. involvement they shifted to larger attacks. In 1972, the number of attacks remained about the

[7] *Counterbattery fire* attempts to identify the location of attacking mortars or rockets (e.g., using radar tracking of incoming shells), then uses that information to direct artillery or mortar fire onto the enemy positions.

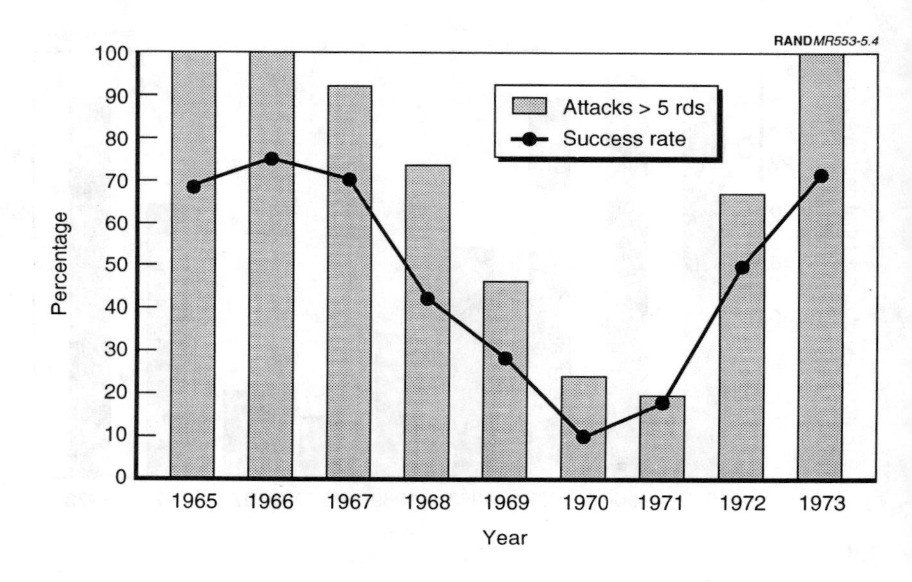

Figure 5.4—Attack Size Versus Success Rate

same as that for the preceding year, but 67 percent of the attacks fired over 5 rounds, more than doubling the success rate. There were only seven attacks in 1973; in each of them, more than 5 rounds were fired.

Figure 5.5 shows aircraft losses to ground attack, by service. The Air Force lost the most fixed-wing aircraft, whereas the Army lost the greatest number of helicopters. Indeed, Army aircraft losses are roughly twice Air Force losses, probably owing to the nature of Army flying: Most Army aircraft were based at vulnerable forward fields, and some helicopters were likely caught on the ground at landing zones during airmobile operations.

Figure 5.6 shows the breakdown of USAF aircraft destroyed by ground attacks. Lost aircraft types span the spectrum from single-engine light utility aircraft, such as the O-1, to much more expensive and sophisticated F-4s. Note that high-value aircraft, such as KC-135s, B-52s, AC-130s, and F-105Gs, were based in Thailand and Guam, where the ground threat was lower or nonexistent.

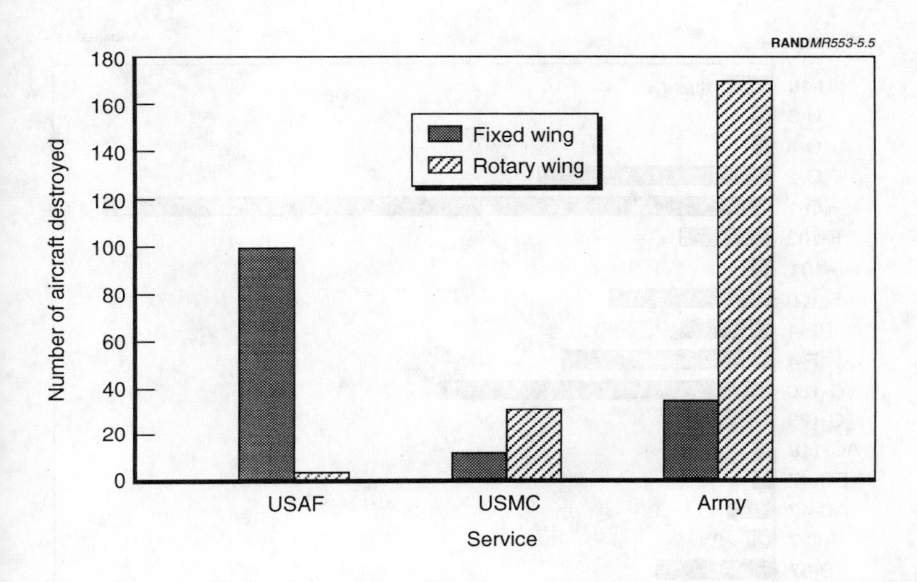

Figure 5.5—Aircraft Losses to Ground Attacks, by Service

DISCUSSION OF SELECTED ATTACKS IN THE REPUBLIC OF VIETNAM

Basic details on each of the 475 attacks against MOBs in the Republic of Vietnam, designated in Figure 5.7, can be found in Appendix B. This section briefly discusses three of these attacks, which are representative of the challenges air-base-defense planners faced.

Air base defense was a low priority for USAF forces in Vietnam until late 1964. For example, in fall 1964, Tan Son Nhut had only six Security Police vehicles to patrol a 16-mile perimeter. Most bases had little perimeter fencing, and the RVNAF was notorious for its poor control over base access. When the war heated up in 1964, these deficiencies made USAF bases an attractive target for Viet Cong forces.

Following the Gulf of Tonkin incident in August 1964, USAF commanders became concerned about the potential for VC/NVA retalia-

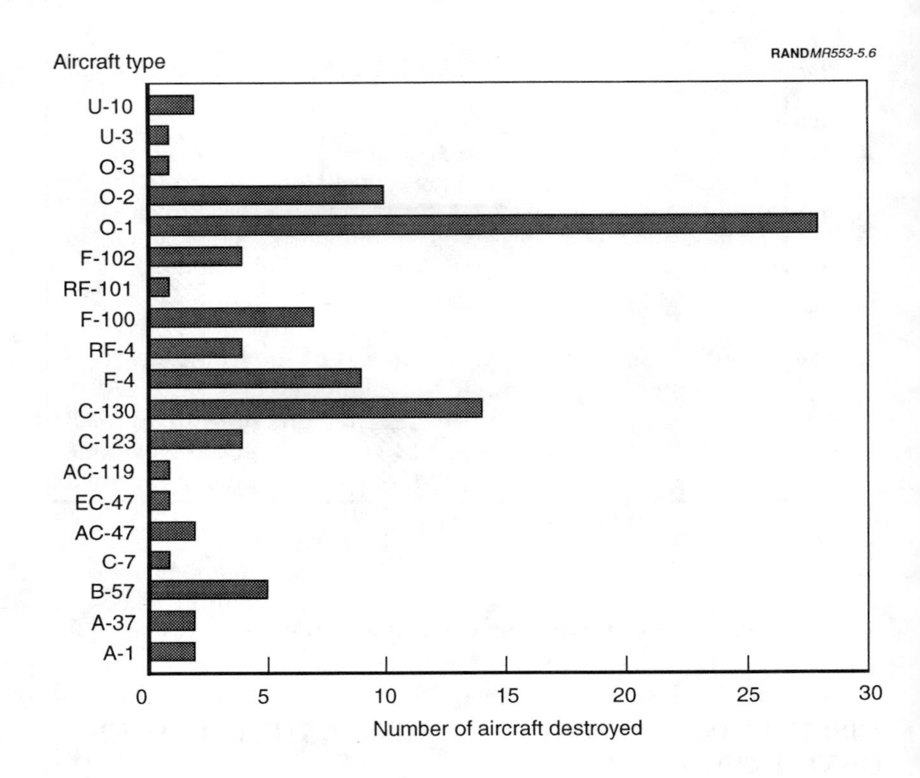

Figure 5.6—USAF Losses to Ground Attack, by Aircraft Type

tion against USAF bases. They requested that the RVNAF increase base security forces, and the 2nd Air Division commander ordered one of the two B-57 squadrons, parked wing tip to wing tip at Bien Hoa, evacuated to the Philippines. The aircraft departed Bien Hoa nine days before the Viet Cong attacked.[8]

The November 1, 1964, Viet Cong attack against Bien Hoa AFB demonstrated what low-technology forces can achieve if they have good intelligence, mission-planning, and weapon skills (see cover

[8]USAF, 1969a, p. 27; Fox, 1979, pp. 14–16.

RANDMR553-5.7

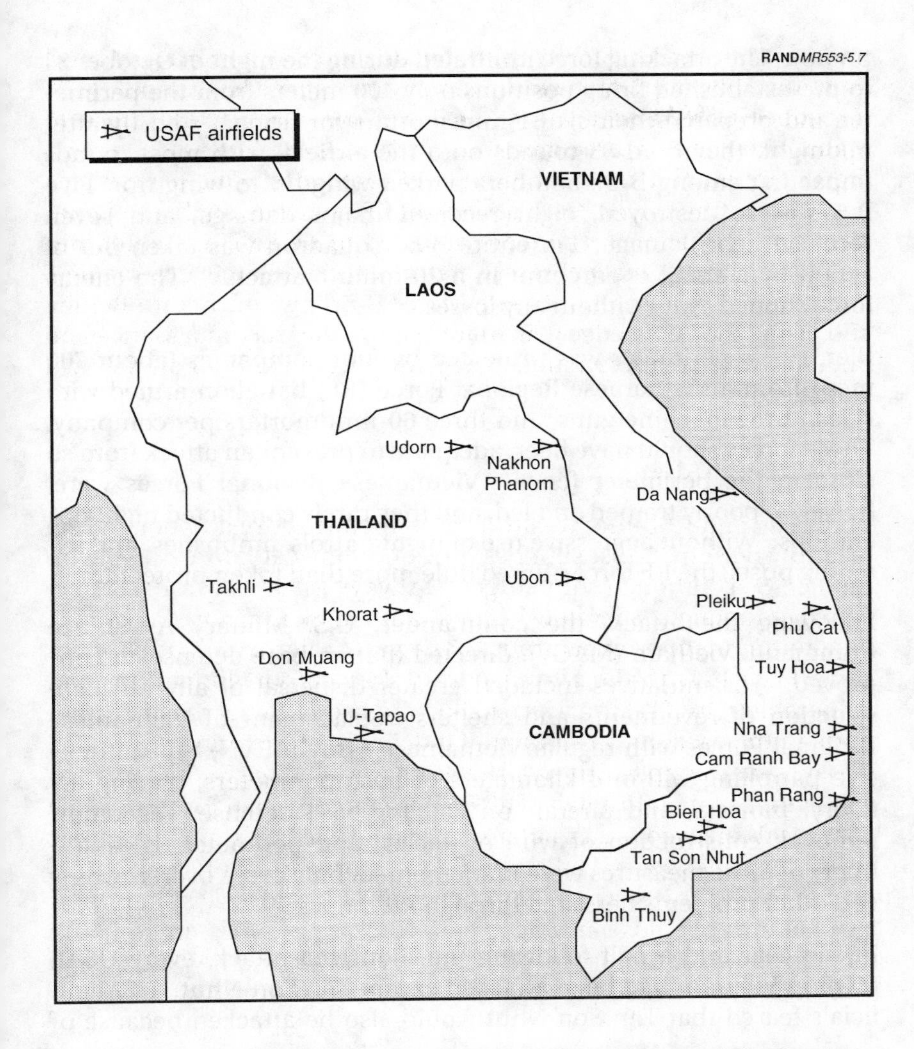

Figure 5.7—USAF Main Operating Bases in Vietnam and Thailand, 1964–1973

photo). The attacking force infiltrated during the night of October 31 to pre-established firing positions only 400 meters from the perimeter and prepared their six 81-mm mortars for firing.[9] Shortly after midnight, they fired 83 rounds onto the airfield, with most rounds impacting among B-57 bombers parked wing tip to wing tip. Five B-57s were destroyed, eight received major damage, and seven received light damage. An entire B-57 squadron was taken out of action by a small enemy unit in a 20-minute attack.[10] The enemy force slipped away without any losses.

Bien Hoa's perimeter was protected by four companies (about 700 men) from a Vietnamese Regional Force (RF) battalion armed with rifles, three machine guns, and three 60-mm mortars per company. These forces should have been adequate to prevent an attack from so close to the perimeter fence. Vietnamese Regional Forces were, however, poorly trained and led, and they rarely conducted night operations. Without aggressive use of night patrols, ambushes, and listening posts, the RF force offered little more than token protection.

Following the attack, the commander, U.S. Military Assistance Command, Vietnam (MACV), directed that air base defenses be improved. His initiatives included greater dispersal of aircraft; construction of revetments and shelters; replacement of Vietnamese Regional Forces with regular Vietnamese Army (ARVN) units; intensive patrolling within 4 kilometers of base perimeters; placing artillery, mortars, and aircraft on call for base defense; vegetation removal; construction of wire obstacles; and perimeter lighting.[11] Many of these measures were implemented, but severe overcrowding and other problems persisted throughout the war.

In the year and a half following the Bien Hoa attack, every USAF MOB in Vietnam had been attacked except Tan Son Nhut. USAF officials feared that Tan Son Nhut would also be attacked because of

---

[9]USAF (Project CHECO) and other reports all describe the mortars as 81-mm as opposed to Russian or Chinese 82-mm weapons issued to NVA forces. Perhaps the 81-mm mortars were captured or acquired in some other fashion from South Vietnamese forces.

[10]USAF, 1969a, p. 27. Also, two Vietnamese Air Force aircraft were destroyed and another three were damaged.

[11]USAF, 1969a, pp. 29–30.

the psychological value of an attack on this major installation; intelligence reports in late 1964 indicated that the VC were planning such an attack.[12] On April 13, 1966, Tan Son Nhut was attacked with mortars and recoilless rifles. The barrage, which lasted 13 minutes, dropped 245 rounds on the base. Two aircraft were destroyed and 62 were damaged (see photo plates). Additional losses included 34 vehicles destroyed or damaged, one 420,000-gallon fuel tank destroyed, and minor runway damage.[13] The Project CHECO report on the attack describes the damage as follows:

> The Viet Cong struck with deadly accuracy. Part of the heavy barrage landed in a fuel storage area, and one tank of fuel exploded in flames. The flames soared hundreds of feet into the night sky, joining flares dropped by Air Force flareships in lighting the area. Air Force firemen quickly fought to contain the blazing fuel stores and extinguished fires at other points on the base. The pinpoint accuracy of the barrage caused considerable damage to aircraft at Tan Son Nhut. In some cases there were direct hits, in other instances the mortar shells landed directly in front of the revetted aircraft and many small pieces of shrapnel penetrated the aircraft.[14]

There is no escaping the fact that this was one of the most destructive enemy attacks of the war. Nevertheless, it could have been worse if USAF leaders had not taken the Bien Hoa attack seriously and implemented numerous corrective actions. Tan Son Nhut improvements included construction of aircraft revetments and wire barriers around high-value areas; installation of mines and trip flares; and assignment of ARVN units to assist the Regional Forces in providing exterior perimeter defense. Regional Force units also were more active and had established night ambushes on likely approaches.

---

[12]USAF leaders had predicted the Viet Cong motivation correctly. Viet Cong prisoners captured in the attack stated that their objective was to destroy aircraft "to prove the Viet Cong were winning and to heighten the morale of VC soldiers and cadres." See Carl Berger, ed., *The United States Air Force in Southeast Asia, 1961–1973*, Washington, D.C.: U.S. Air Force Office of History, 1984, p. 263.

[13]USAF, *Attack Against Tan Son Nhut: Project CHECO Southeast Asia Report*, Hickam AFB, Hawaii: Headquarters, Pacific Air Forces, 1966, pp.1, 8; USAF, 1969a, p. 30.

[14]USAF, 1966, pp. 7–8.

Almost concurrent with the beginning of the VC barrage, a small Regional Force element detected and ambushed part of the attacking force. If more ambush positions had been manned, the VC force might have been detected prior to the attack. USAF guard dogs performed well, detecting activity along the southwest and west perimeters. Small-arms fire was exchanged, and the attempted penetration was defeated. Also, U.S. helicopter gunships and flareships were launched within 20 minutes of the initial assault. The gunships fired on muzzle flashes in the area of the VC mortar positions, silencing them. One major disappointment for the defenders was the failure of their new counterbattery radar to identify any of the enemy firing positions because they were under the radar's minimum range.[15]

In contrast to other USAF MOBs, at Da Nang, USAF Security Police were not primarily responsible for base defense. Da Nang was the command and logistics center for the III Marine Amphibious Force and for I Corps. As such, it was a high-priority area for the Marine Corps, who were responsible for base defense. USAF forces manned one-tenth of the perimeter and flew reconnaissance and/or gunship patrols.[16]

The VC and NVA tested these defenses and demonstrated a significant new capability on February 27, 1967, when they attacked the base with Soviet 140-mm rockets. The barrage dropped 64 rounds on Da Nang in less than 1 minute, damaging 13 aircraft.[17]

After the attack, 134 rocket-firing positions were discovered 8 kilometers southwest of the base. The range, light weight, and simplicity of the rocket led to many more rocket attacks against U.S. bases. Deployment of 102-, 122-, and 140-mm rockets permitted the VC and NVA to attack bases from a maximum range of 11 kilometers. Following this attack, USAF and other friendly forces had to expand their tactical area of operations (TAOR) to include this rocket belt. Army and Air Force forward air controllers flew numerous daytime

---

[15]USAF, 1966, p. 9; USAF, 1969a, pp. 31–32.

[16]USAF, *Defense of Da Nang: Project CHECO Southeast Asia Report,* Hickam AFB, Hawaii: Headquarters, Pacific Air Forces, 1969b, p. 1.

[17]USAF, 1969b, pp. 32–34.

surveillance missions in the Da Nang area. The Marines, however, were the only service that flew patrols specifically targeted at the rocket belt.

Using borrowed Army OH-6A helicopters, the Marines flew three patrols per day at treetop level and often spotted enemy forces. When rocket sites were detected, they were attacked with artillery or by infantry. Night patrols were rarely flown, because the aircraft lacked night-surveillance equipment. Also at Da Nang, the Marines constructed an anti-infiltration fence stretching 48 kilometers along the maximum-range curve of a 122-mm rocket fired at the base. The ARVN section of the fence was composed of barbed-wire fences, observation towers, and bunkers. The Marine section consisted of a 500-meter-wide open area cut through the forests and grasslands. In addition to barbed wire, there was a chain of sensors that could detect personnel or vehicle movement. The First Marine Division considered this system reliable and reported that it forced enemy troops to make wide detours to sensorless sections of the fence. Enemy forces that failed to make the detour were routinely detected and engaged by artillery or ground forces.

While the fence and sensors undoubtedly increased the Marines' interception rate, they by no means stopped rocket attacks on Da Nang. The area between the fence and allied facilities was densely populated. During daylight hours there was considerable movement of people and products into and out of this zone. No practical means was available for security forces to monitor or inspect all the traffic. Thus, enemy forces were able to penetrate into the rocket-launching belt routinely; by February 1968, they had fired a total of 297 rockets into the air base, causing $110 million in damage. In the words of the commander of the 366th Tactical Fighter Wing stationed at Da Nang, these attacks "significantly interfered with our combat mission."[18]

Perhaps the most effective countermeasure to the rocket attacks was a passive one. By April 1969 the Air Force had built 98 steel arch shelters at Da Nang. These shelters were covered with 15 inches of concrete and proved impenetrable by 140-mm rockets. (See photo plates.)

---

[18]USAF, 1969b, pp. 7–8, 13–15.

AIR BASE DEFENSE IN THAILAND—FIVE ATTACKS

Background

USAF records document five attacks on U.S. bases in Thailand. Because these have not been previously reported in an unclassified publication,[19] this section provides a brief summary and discussion of each attack.

U.S. units were stationed at Udorn, Ubon, Khorat, Don Muang, Takhli, U-Tapao, and Nakhon Phanom Royal Thai Air Force bases (see Figure 5.7).[20] By December 1967, 505 USAF aircraft were stationed in Thailand.[21] Officially, USAF security responsibilities were limited to close-in protection of their own resources on the Royal Thai Air Force bases. The Royal Thai Air Force, Army, and various police organizations were responsible for detecting and preventing both standoff attacks and attempts to penetrate the base perimeters. It became clear, however, that Thai forces were not up to the task, and USAF Security Police ultimately became responsible for the planning, command, and execution of defensive operations. USAF intelligence personnel viewed North Vietnamese infantry or Thai Communist insurgents armed with mortars, recoilless rifles, and rockets as the primary threats to air bases in Thailand. Accordingly,

> the enemy was to be denied unhindered operational access to all areas within a 10,000 meter radius of each base. The most significant area to be denied was the 5,000 to 10,000 meter belt, where the enemy could employ 81-, 82- and 120-mm mortars, and 122- and 140-mm rockets. That was the area from which they could hit each base with a resultant high level of damage and, due to the long range, be almost undetectable. Observation posts in the higher threat areas, flareships and gunships on alert, free-fire zones around the bases, and forces readily available for prompt and decisive countermeasure deployments to conduct ambushes and of-

---

[19]The Air Force declassified this Project CHECO report in August 1994. See footnote 21 for the full citation.

[20]There are no written reports of ground attacks on Khorat, Don Muang, Takhli, or Nakhon Phanom.

[21]USAF, *Base Defense in Thailand: Project CHECO Southeast Asia Report,* Hickam AFB, Hawaii: Headquarters, Pacific Air Forces, February 18, 1973, p. 2.

fensive ground action against enemy training areas and hide-outs were envisioned.[22]

Coordination problems with the various Thai governmental agencies, shortages in equipment and personnel, and insufficient training opportunities limited the implementation of these sound defensive plans. The most intractable problem was the shortage of trained personnel for the manpower-intensive missions of patrolling, manning fighting positions, and providing quick-reaction forces. Thai sensitivities about the size and visibility of the U.S. presence, the U.S. government's desire to limit the size of ground combat elements outside of Vietnam, and demands for Security Police in Vietnam combined to produce serious manpower shortfalls at all bases.

To fulfill the manpower requirements, the U.S. Military Assistance Command in Thailand and the Thai government signed a Memorandum of Understanding in 1966. This agreement created a Thai contract guard force, the Thai Security Guard Regiment, to augment air base defenses. Operational control of these units was given to each USAF base commander.[23] The Thai guards provided most of the manpower and generally performed well under USAF leadership. Thai bases, however, never received the resources necessary to provide security equivalent to that given USAF MOBs in Vietnam. Throughout the war, USAF bases in Thailand lacked sufficient perimeter fencing, lighting, observation towers, and defensive fighting positions. These shortfalls made it possible for NVA sappers to penetrate base perimeters on at least five occasions.

Udorn—Attack 1

The first of these attacks (Attack 1) was against Udorn Royal Thai Air Base (RTAB) on July 26, 1968. There was no warning of an impending attack; base security forces were at their routine posture. Special security was being provided, however, for a C-141 transport on call for a priority evacuation mission. The extra security included placing a close-in sentry next to the C-141, positioning additional sentries

---

[22]USAF, *Attack on Udorn (July 26, 1968): Project CHECO Southeast Asia Report*, Hickam AFB, Hawaii: Headquarters, Pacific Air Forces, December 17,1968, pp. 3–4.

[23]USAF, 1968, pp. 8–9.

between the taxiway and perimeter, and posting a special quick-reaction team nearby. At 10:25 p.m., approximately 25 attackers—from four separate locations—opened fire with automatic weapons against the northwest corner of the base. It appears that this attack was a diversion, because, at the same time, several sappers attempted to reach the C-141. The close-in guard killed one sapper under the tail of the aircraft and another sapper 20 yards away. A third intruder fired his AK-47 rifle into the area around the C-141, which appears to have caused a fuel leak from one of the aircraft's engines. The sapper then threw an explosive charge under the aircraft and another under a Mobile Power Unit. The first charge ignited fuel that was pouring from the damaged engine. The sapper then ran down the length of the taxiway toward two F-4D aircraft. These aircraft were undergoing maintenance and did not have any special security. The sapper threw an explosive charge into the back of a Security Police truck and another into the tailpipe of an F-4. The charge in the F-4 failed to detonate; the sapper returned and placed another. The second charge detonated; the sapper then ran into the grass and escaped.

An HH-43 helicopter equipped for fire suppression and ground fire-fighting units were able to stop the C-141 fire, but the HH-43 was damaged by small-arms fire. Quick-reaction forces responded within 2 minutes of the original attack and engaged the remaining attackers with small-arms fire. The attackers then retreated. The attack caused heavy damage to the C-141, moderate damage to the F-4, and light damage to the HH-43. Light damage was done to four USAF vehicles, a power unit, and a light unit.[24]

Ubon—Attacks 2 and 3

The next attack (Attack 2) came one year and two days later, at Ubon RTAF. At 1:30 a.m. on July 28, 1969, a Security Policeman and his dog were wounded when they detected three sappers attempting to leave the base. Thirty minutes later, five explosions damaged 2 C-47 aircraft and a van. Five additional dud charges were also found. The chief of Security Police at Ubon identified the major deficiencies that

---

[24]USAF, 1968, pp. 16–24.

made this attack possible as the failure to use available night-vision devices, inadequate perimeter vegetation control, and poor dog-handler training.[25]

Ubon was attacked again in 1970 (Attack 3). A local villager reported seeing 16 armed Vietnamese 3 kilometers from the base at 10:30 p.m. on January 11. Subsequently, the base was put on alert, with 363 security personnel on duty. At 2:01 a.m. on January 12, a sentry detected and fired on an enemy sapper 30 feet inside the perimeter fence. The sentry was reinforced by the sector alert team,[26] and the sapper was joined by five other attackers. In the resulting firefight, five of the sappers were killed and the attack stopped; 35 satchel charges were found with the bodies. The USAF assessment cited good intelligence, training, command and control, and a fast response by the sector alert team as the keys to the detection and defeat of the enemy.[27]

U-Tapao—Attack 4

It was two years before a base in Thailand was attacked again; this time the attackers chose U-Tapao, considered the safest base in the country (Attack 4). At 2:22 a.m. on January 10, 1972, three sappers penetrated the base perimeter and approached to within a few hundred yards of a B-52 aircraft when they were detected by a guard dog. The sappers fired on the sentry without effect. One sapper disappeared, and the other two ran toward the B-52s. Thai Guards saw, but failed to stop, the sappers before they threw a grenade and four satchel charges into 3 B-52 revetments, causing moderate damage to 1 B-52 and minor damage to 2 others.[28] The Air Force Office of Special Investigations' (OSI) assessment of the attack observed that

> the relative degree of success or failure of the U-Tapao attack depends on who is making the assessment. From the communist standpoint, they infiltrated three intruders into a heavily defended

---

[25]USAF, 1973, pp. 6, 9.

[26]The *sector alert team* was a small, quick-reaction force whose mission was to reinforce defensive positions under attack.

[27]USAF, 1973, pp. 9–10.

[28]USAF, 1973, pp. 10–11.

> U.S. position, damaged three expensive U.S. aircraft, and recovered two of the attackers. The loss of only one man, when measured against the satisfaction and propaganda value derived from such an effort, clearly marks the success of the mission. From the American side, the early detection of the intruders and their failure to significantly affect U.S. combat posture makes the attack a failure. Regardless of which viewpoint is accepted, the U-Tapao attack serves to reaffirm the contention that small groups of well trained, dedicated individuals can penetrate U.S. tenanted installations in Thailand.[29]

Following this attack, U-Tapao Security Police made several adjustments. Recognizing that they lacked the manpower to monitor the long perimeter of this large base, they focused on improving secondary and close-in defenses. Dog patrols were concentrated in the middle defenses, and ambushes were laid along likely avenues of approach between the perimeter and aircraft revetments. Close-in defenses were strengthened by assigning one guard per aircraft.

Ubon—Attack 5

The final attack (Attack 5) occurred six months later, at Ubon, the target of choice in Thailand. On June 1, 1972, the OSI unit at Ubon received a report that 12 Vietnamese expatriates living in the Ubon area had recently returned from a trip to North Vietnam, where they received sapper training. A few minutes after midnight on the night of June 3/4, Thai provincial police reported seeing a man just inside the perimeter fence running toward AC-130 revetments 50 yards away. The police exchanged fire with the sapper, and he was killed. Eight satchel charges were found on the body. Additional attempts to penetrate the perimeter were detected that night. No other penetrations succeeded, and there was no damage to aircraft.

After this attack, there were three separate contacts with the remainder of the sapper force in the vicinity of the Thai-Laotian border. In the final engagement, Royal Laotian Army forces killed two of the sappers and identified them as regular NVA soldiers.[30]

---

[29]USAF, 1973, p. 13.

[30]USAF, 1973, pp. 13–15.

AIR-BASE-DEFENSE CONCEPT OF OPERATIONS

Layered Defense

USAF bases in Vietnam and Thailand developed layered defenses against both standoff and penetrating threats. The *first layer* was the general vicinity around the base, in which friendly ground forces, police, and intelligence sources provided some early warning and occasional engagement of threat forces. The base perimeter fence was the *second layer*, providing the primary barrier and opportunity to detect sappers. Observation towers and bunkers were widely used in this layer. The defensive concept was one of

> firmly fixing and engaging the attacking force to prevent its access to the base. This is accomplished through the use of obstacles, barbed wire, minefields and trip flares to delay, harass, and channel enemy forces into established fields of fire. It relies upon superior firepower from prepared defensive positions (machine gun bunkers and mortar positions). The firepower available at USAF bases [in the Republic of Vietnam] since the 1968 Tet Offensive is overwhelming. All bases have 50-calibre machine guns, both mobile and fixed M-60 machine gun positions, 81-mm mortar and sufficient quantities of M-16 rifles for both Security Police reserve augmentees and mass arming of base personnel.[31]

The *third layer* was composed of roving security alert teams, sentries, and patrol dogs to detect penetrations of the perimeter. In Vietnam, these personnel were supplemented by mobile 12-man Quick-Reaction Teams, mounted in either M-113 armored personnel carriers or M-706 armored cars. Jeeps, trucks, and assorted other vehicles were used when the armored vehicles were not available. Finally, high-value sites were protected with defensive positions, patrols, and sentries. For example, one sentry was assigned for every eight aircraft in daylight hours and one for every four at night. B-52 and KC-135 aircraft received double coverage.[32]

---

[31]USAF, *7th Air Force Local Base Defense Operations (July 1965–December 1968): Project CHECO Southeast Asia Report*, Hickam AFB, Hawaii: Headquarters, Pacific Air Forces, 1969a, p. 16.

[32]USAF, 1973, p. 35.

Close Air Support

Close air support was provided by AC-47 and helicopter gunships. Base rescue helicopters were used for airborne surveillance as early as November 1964. At Nakhon Phanom, the Security Police used HH-53 rescue helicopters to conduct twice-nightly reconnaissance flights out to 16 kilometers from the perimeter. Each of these flights typically lasted 3 hours, resulting in 6 hours of airborne surveillance nightly. The reports do not indicate whether night-surveillance devices were used or if flares were dropped; it is not clear what could be seen without these aids during these patrols. Clear nights with good moonlight offered much better visibility, but the VC and NVA, like the SAS 20 years earlier, tended to launch attacks on nights with little or no lunar illumination.[33] Observation flights during the day proved useful in preventing standoff attacks, because they often detected preparation of rocket-launch sites (e.g., pits, mounds, bamboo launch platforms). Eventually, most bases in Vietnam made arrangements with the Army or Marine Corps to keep helicopter gunships on a 3-minute alert for base defense. Some bases used airborne alerts during the peak threat period (10 p.m.–3 a.m.).

Counterbattery Fire

Counterbattery fire was one option that had the potential to make standoff attacks less attractive, and perhaps as risky as penetrating attacks. A Project CHECO report summarizes the process:

> Standoff mortar/rocket attack locations can be detected after firing commences by plotting azimuths reported by tower observers on an M-5 plotting board at Combat Security Control (CSC). This system, called the "flash base system," has been used effectively at Bien Hoa Air Base, and security police observers in strategically located towers have been able to consistently give base personnel 16-to-20-second warnings of an impending attack by activating the base siren system from switches located in the towers. Azimuth sightings from the direction of the rocket flash are then reported to

---

[33]For example, at Cam Ranh Bay, where during the first nine months of 1970, 50 percent of the attacks occurred when lunar illumination was below 25 percent. See USAF, *Attack on Cam Ranh (25 August 1971): Project CHECO Southeast Asia Report*, Hickam AFB, Hawaii: Headquarters, Pacific Air Forces, December 15, 1971b, pp. 9–10.

> CSC by radio and the source of fire can be plotted within 100-meter accuracy within 20 seconds. This information is relayed to artillery units on base, and range can be calculated and fire returned within two-to-three minutes after rocket flash is observed. Artillery fire is effective within a 100-meter radius and firing sequence is 100 meters over, under, both sides, and then on target. This is designed to hit the enemy before he can withdraw. Of significant note, although Bien Hoa Air Base has counter-mortar radar, attack warnings in 3 standoff mortar/rocket attacks during the period 1 February to 17 June 1969 were initiated by Security Police tower observers.[34]

As was often the case in the Vietnam War, a promising military solution was significantly constrained by other factors. For Bien Hoa, most attacks were launched near villages between two ARVN areas of operation. In such situations, MACV rules of engagement required that the Bien Hoa base commander request clearance to fire from both the Vietnamese province chief and the ARVN commander. This cumbersome process made effective counterfire impossible. Only 25 percent of the requests were approved; and even when approved, the process took so long that the attackers were typically long gone before the counterbattery fire commenced.

Control of Standoff Footprint

The most effective means to deter and prevent standoff attacks was to control the standoff footprint around the base. The longest-range threat was 122-mm rockets, reaching out 11 kilometers. To defeat the standoff threat, therefore, required controlling territory extending this distance from the base perimeter, typically with a total area of over 200 square miles. Although ground forces were not assigned the mission of controlling this footprint, in some cases high-quality friendly ground forces did patrol in the vicinity of air bases. For example, at Phu Cat and Nha Trang, aggressive patrolling and other operations by Republic of Korea forces have been given credit for reducing standoff attacks. At Da Nang, both air and ground forces monitored the rocket belt quite extensively.

[34]USAF, 1969a, pp. 12–13.

Passive-Defense Measures

Crowding at some bases limited the use of passive-defense measures, such as revetments, shelters, and aircraft dispersal. Passive defense received new attention when the VC/NVA used Soviet rockets against Da Nang in a February 1967 attack. Rockets gave enemy forces a long-range and potent weapon that could be countered only by artillery or aircraft. Reacting to this new threat, the USAF tested a number of different approaches, settling on concrete revetment roofs or new shelters for its fighter aircraft (see photo plates). In March 1969, one of the roofed revetments suffered a direct hit by a 140-mm rocket. The aircraft inside went unharmed (see photo plates).[35] In contrast, open revetments appeared to offer limited protection. For example, in a major mortar-and-recoilless-rifle attack on Tan Son Nhut in 1966, 23 of the 61 damaged aircraft were in revetments. Another 39 aircraft in revetments were not harmed; reports do not indicate whether such structures were inside or outside of the impact area. Revetments certainly helped contain fires and secondary explosions but provided no protection from direct hits. Typically, two tactical aircraft were parked in each revetment; a direct hit by a single round could destroy both. Also, revetments opened toward one another across a taxiway; rounds landing in the taxiway could damage multiple aircraft. (See photo plates.)

ATTACK TACTICS

The selected attacks in Vietnam and Thailand have been described and base-defense concepts of operations discussed to give the reader a sense of VC/NVA base-attack tactics. Speaking more generally, we can draw three conclusions about these tactics. First, the VC and NVA chose standoff weapons for 96 percent of their attacks. Second, the average attack was quite small, with fewer than 10 rounds fired. Finally, it appears that they had a preference for Sunday attacks. Figures 5.8 through 5.10 illustrate these points.

---

[35]USAF, 1969b, pp. 17–23.

Sapper Attacks

Sapper attacks are prominent in Vietnam war images, but independent sapper attacks did little damage on air bases. Only 21 pure sapper attacks occurred against MOBs in the entire war.[36] Twelve did no significant damage, one destroyed $400,000 of munitions, and the remaining eight destroyed 5 aircraft and damaged an additional 21. However, they are an important tactic and were very successful against non-USAF facilities in Vietnam—and could be used in the future.

It is interesting that all five attacks against MOBs in Thailand were by sappers; perhaps the demands of carrying standoff weapons and munitions across Laos and Thailand proved too much for North Vietnamese logistics.[37] Alternatively, the NVA may have recognized that the bases in Thailand were more vulnerable to penetration. In Vietnam proper, there were eight combined sapper-and-standoff attacks, resulting in 8 aircraft destroyed, 49 damaged, and 460,000 gallons of fuel, 2.25 million gallons of fuel-storage capacity, and 6,000 tons of munitions destroyed. It is not clear whether the indirect-fire support for these combined attacks caused most of the destruction or whether it was a diversion that helped the sappers get to their targets (as RAF bombing raids assisted SAS sappers in North Africa).

Figure 5.8 breaks down sapper attacks by base. Every base in Vietnam received at least one sapper attack. Cam Ranh Bay was attacked seven times; Phu Cat and Phan Rang tied for second with three attacks each.

---

[36]Attempted penetrations of bases probably number in the hundreds, but their number is not given in any of the Project CHECO reports or in Fox's book. It appears that, for an incident to be included in the official statistics, a sapper had to penetrate the perimeter or be killed in the attempt. Given the thousands of incidents in which nervous perimeter guards fired at animals, shadows, wind-driven foliage, and other false targets, the number of such "incidents" would be a poor measure of actual penetration attempts.

[37]There is one unconfirmed report that a 21-man team with mortars infiltrated on foot all the way to Korat AFB in Thailand in 1972 or 1973. This attack was reportedly aborted when the forward observers—who had penetrated the perimeter—were detected and one was captured.

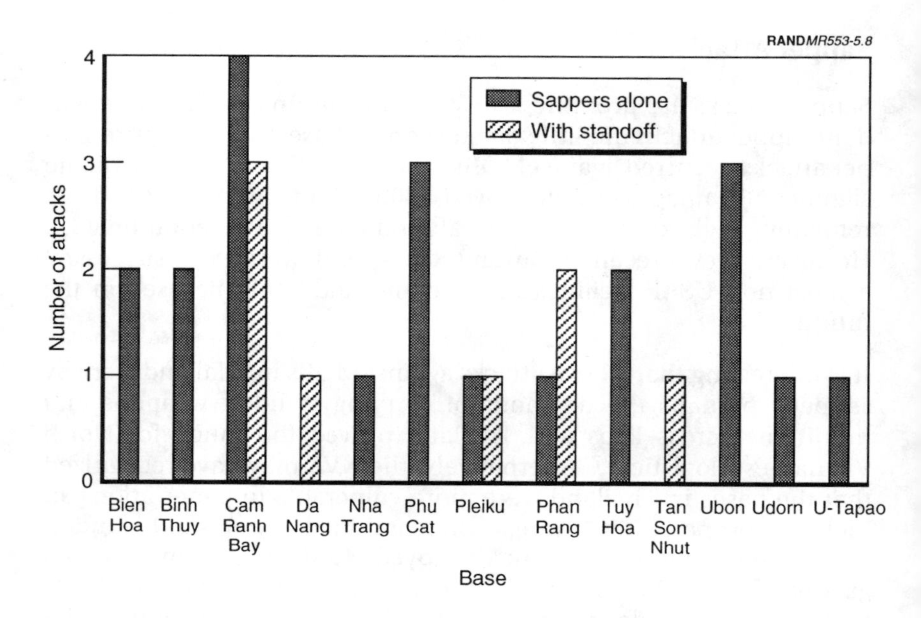

Figure 5.8—Sapper Attacks, by Base

Smallness of Standoff Attacks

Figure 5.9 shows that most standoff attacks were quite small, with almost 300 attacks in which fewer than 10 rounds each were fired. At the other end of the spectrum, only five attacks fired over 100 rounds. The most common number of rounds fired was only three, the choice in 58 attacks.

Sunday Attacks

The daily distribution of attacks is shown in Figure 5.10. Given the proximity of pleasurable distractions and VC knowledge about the leisure-time activities of U.S. servicemen, it seems reasonable that the VC might have focused attacks on weekends. A random distribution would be 69 attacks per day. Although weekends do not vary greatly from the norm, Sundays had 89 attacks, almost 30 percent higher than predicted.

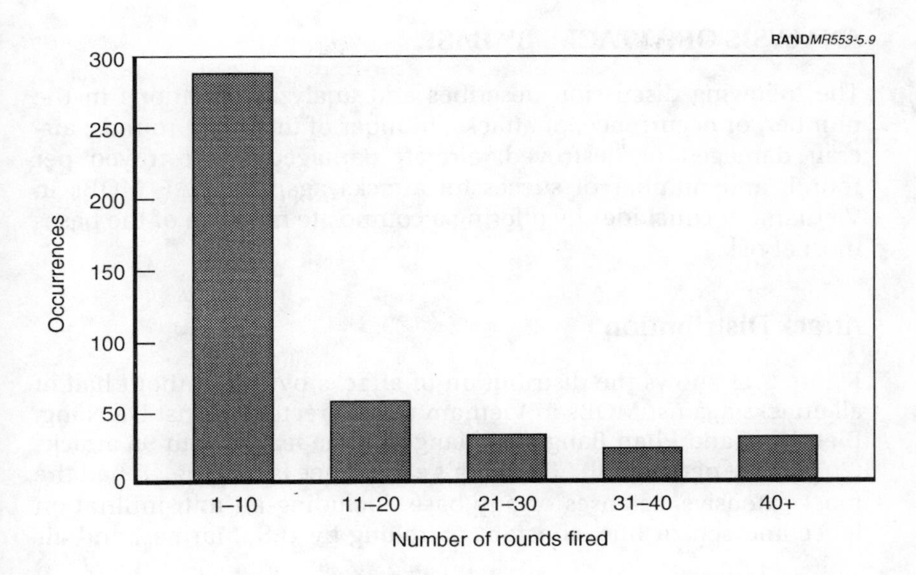

Figure 5.9—Number of Rounds Fired in Standoff Attacks

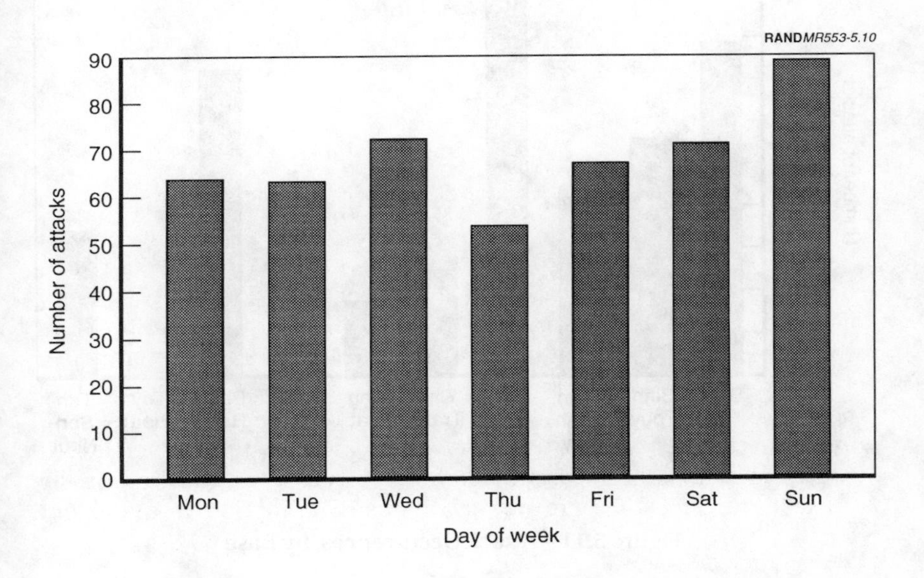

Figure 5.10—VC/NVA Timing of Attacks Against USAF MOBs

ANALYSIS OF ATTACKS BY BASE

The following discussion describes and analyzes variations in the number, or occurrence, of attacks; number of incoming rounds; aircraft damaged or destroyed; aircraft damaged or destroyed per round; and number of successful attacks against USAF MOBs in Vietnam. It concludes by offering a composite measure of the bases most at risk.

Attack Distribution

Figure 5.11 shows the distribution of attacks by base. About half of all attacks against MOBs in Vietnam were directed against Da Nang, Bien Hoa, and Phan Rang. Da Nang was the leader with 95 attacks (20 percent of the total). Da Nang's experience is striking. It had the most extensive defenses of any base, including an anti-infiltration fence and sensor line, constant patrolling by U.S. Marines, and air

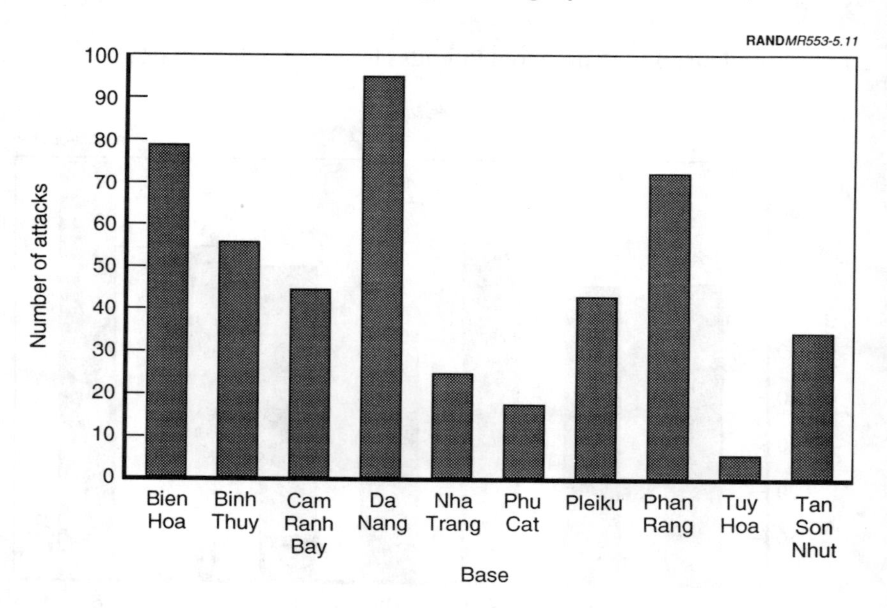

Figure 5.11—Attack Occurrences, by Base

support from three services; yet, it suffered continual attacks right up to the end of the war.

Rounds Fired in Standoff Attacks

Figure 5.12 shows the number of rounds fired in standoff attacks against USAF MOBs in Vietnam. As one might expect, the three air bases attacked most frequently also received the highest number of incoming rounds. Bien Hoa was first on this measure, with almost 1,300 rounds, Da Nang was second with 1,000 rounds, and Phan Rang was third with 700 rounds.

Aircraft Damaged or Destroyed

Figure 5.13 shows the number of aircraft damaged or destroyed at USAF MOBs in Vietnam. Bien Hoa led in aircraft damaged and de-

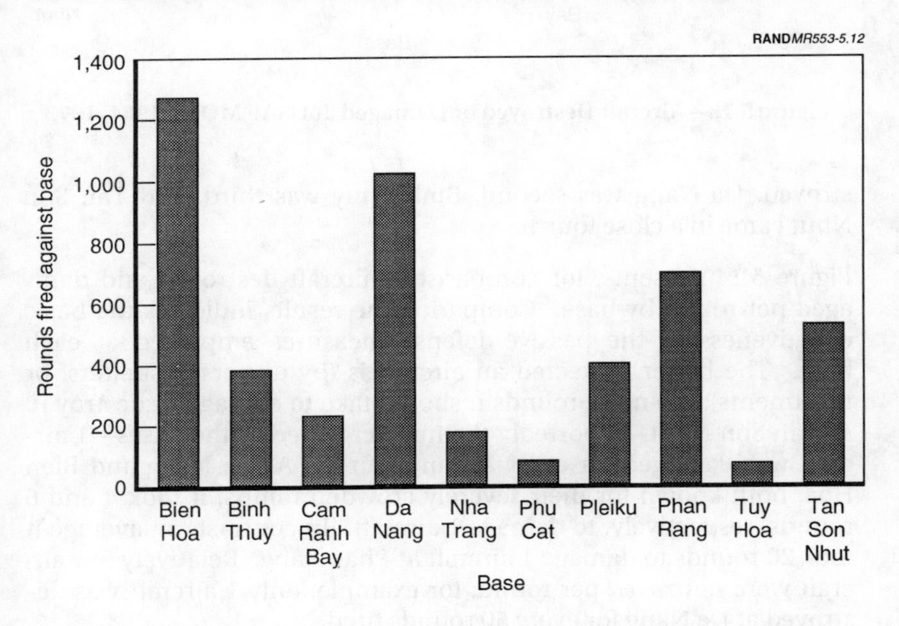

Figure 5.12—Rounds Fired in Standoff Attacks Against USAF MOBs in Vietnam

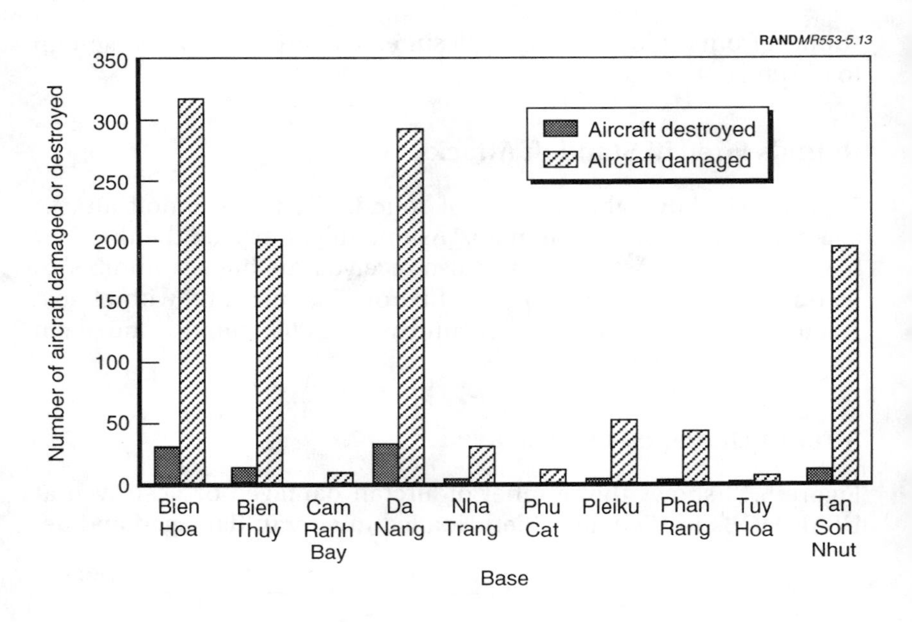

Figure 5.13—Aircraft Destroyed or Damaged at USAF MOBs, 1964–1973

stroyed, Da Nang was second, Binh Thuy was third, and Tan Son Nhut came in a close fourth.

Figure 5.14 presents, for comparison, aircraft destroyed and damaged per round by base. Comparing the results indicates the basic effectiveness of the passive-defense measures employed at each base. The better protected an aircraft is (by dispersal, shelters, or revetments), the more rounds it should take to damage or destroy it. At Tan Son Nhut—reportedly the most crowded of the bases—1 aircraft was damaged for every 3 rounds fired. At Da Nang and Bien Hoa, both known for their severely crowded ramps, it took 4 and 6 rounds, respectively, to damage 1 aircraft. In contrast, on average it took 20 rounds to damage 1 aircraft at Phan Rang. Relatively few aircraft were *destroyed* per round; for example, only 1 aircraft was destroyed at Da Nang for every 50 rounds fired.

Finally, Figure 5.15 shows the percentage of attacks that *succeeded* (destroyed or damaged aircraft), by base. Again, Da Nang's defenses

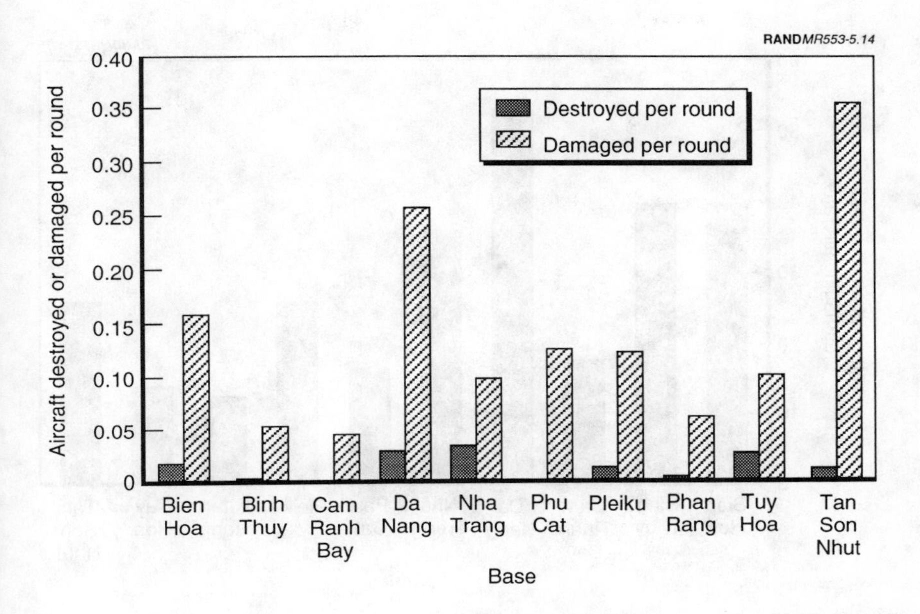

Figure 5.14—Aircraft Destroyed or Damaged per Round in Standoff Attacks Against USAF MOBs in Vietnam

did the worst, with 53 percent of the attacks directed against it succeeding. Bien Hoa and Binh Thuy tied for second, with 39 percent of the attacks against their defenses succeeding. VC/NVA attackers had the lowest success rate against Phu Cat (17 percent), Tuy Hoa (14 percent), and Phan Rang (13 percent).

A crude composite of these various measures (attack number, incoming rounds, aircraft damaged and destroyed, aircraft damaged per round fired, and attack success rate) suggests that aircraft at Da Nang were most at risk, followed closely by those at Bien Hoa. Tan Son Nhut was third, and Binh Thuy was right behind it. The number of attacks and sheer number of rounds fired at Bien Hoa and Da Nang may account for some of their problems. Bien Hoa also suffered from some command-and-control problems in its rear-area

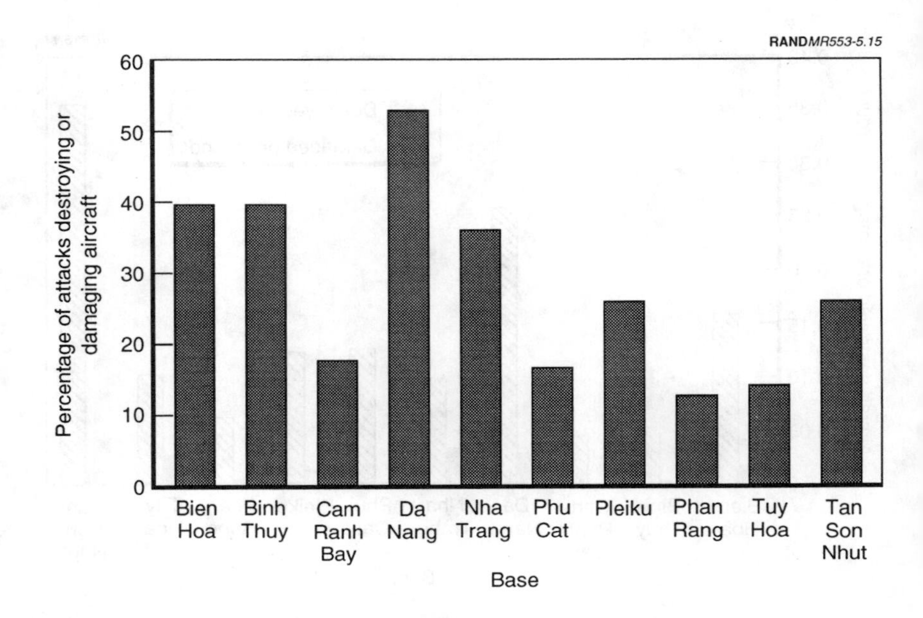

Figure 5.15—Attack Success Rate, by Base

security, because most standoff attacks against it were launched from a position between a nearby village and ARVN positions.[38]

As mentioned above, Tan Son Nhut and Da Nang had reputations for having severely crowded ramps. Was this as significant a factor as it appears to be? Plotting the number of aircraft assigned to each base against the number of aircraft lost per round indicates no relationship (see Figure 5.16).

The base that lost the most aircraft per round (Tan Son Nhut) had only 230 assigned aircraft. In contrast, Nha Trang, with 246 assigned aircraft, lost only 0.10 aircraft per round. It is also strange that Da Nang and Bien Hoa, with 347 and 515 aircraft, respectively, did better than Tan Son Nhut.

---

[38]USAF, 1969a, p. 10.

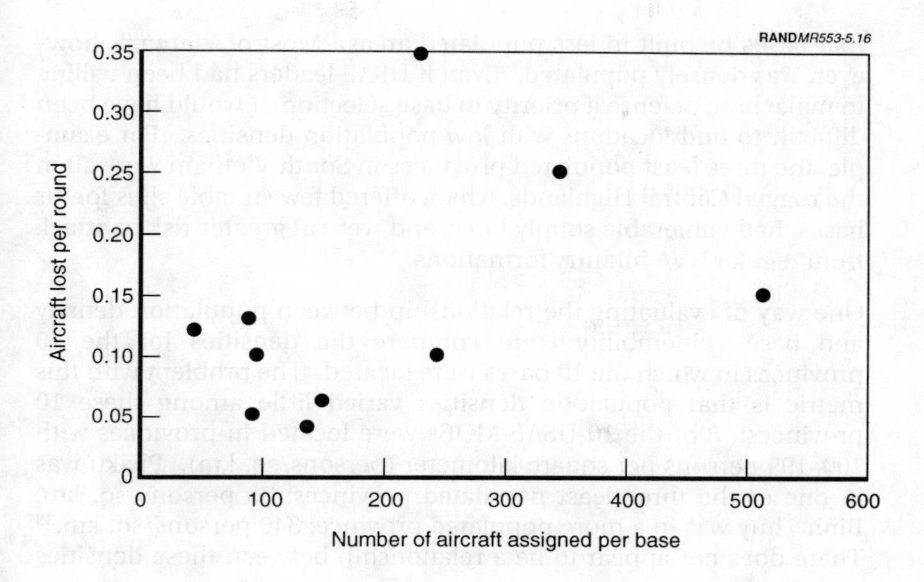

Figure 5.16—Aircraft Losses per Round Versus Number of Aircraft Assigned to Each Base

A better measure of aircraft density would be to compare the number of aircraft assigned and the number of transiting aircraft with square footage of ramp space. Aircraft flowthrough and ramp space data were not available, however. Thus, we are left with the somewhat unsatisfying conclusion that ramp crowding *probably* contributed to aircraft losses.

Population Density

Crowding of another sort—variations in population density—has been suggested by some as a possible explanation for the disparities in the effectiveness of defenses. High densities made it difficult to control the movement of people and goods and enabled the attackers to blend in with the local population. Viet Cong sympathizers living in the area could provide shelter and information about base defenses. Furthermore, the presence of homes and farms bordering the bases limited the use of minefields, flares, and defensive counterfire. There is little doubt that base defenders would have preferred that bases be built in less populated areas. Most of Vietnam, however, was densely populated. Even if USAF leaders had been willing to make base defense a priority in base selection, it would have been difficult to find locations with low population densities. For example, the three least populated provinces in South Vietnam were all in the rugged Central Highlands, which offered few suitable sites for air bases, had vulnerable supply lines, and were at greater risk of attack from regular NVA infantry formations.

One way of evaluating the relationship between population density and base vulnerability is to compare the densities for the 10 provinces in which the 10 bases were located. The problem with this metric is that population densities varied little among these 10 provinces: 8 of the 10 USAF MOBs were located in provinces with 100–199 persons per square kilometer (persons/sq. km). Pleiku was in one of the three least populated provinces: 23 persons/sq. km. Binh Thuy was in a more populated province: 340 persons/sq. km.[39] There does not appear to be a relationship between these densities and attack success rates (see Figure 5.17).

Province-wide densities do not, however, tell us what we really want to know: the densities within a few kilometers of the bases. Unfortunately, this information was not available for all bases. We do know that several of the bases were surrounded by dense local populations. For example, Tan Son Nhut and Da Nang were located adjacent to metropolitan Saigon and Da Nang City, respectively. Homes were built right up to the perimeter fence at Nha Trang; local residents often used the fence as a clothesline. A 1969 study concluded that evacuating a 1-mile security belt around Bien Hoa—one of the most vulnerable bases—would have displaced 14,000 people; a similar belt around Tuy Hoa—one of the safest bases—would have displaced over 16,000 people.[40]

From this evidence, it does not appear that variations in population density were a significant factor in the success rates of air base at-

---

[39]Judith Bannister, *The Population of Vietnam*, Washington, D.C: U.S. Department of Commerce, 1985, Table Two and Map Two.

[40]Fox, 1979, pp. 60–61.

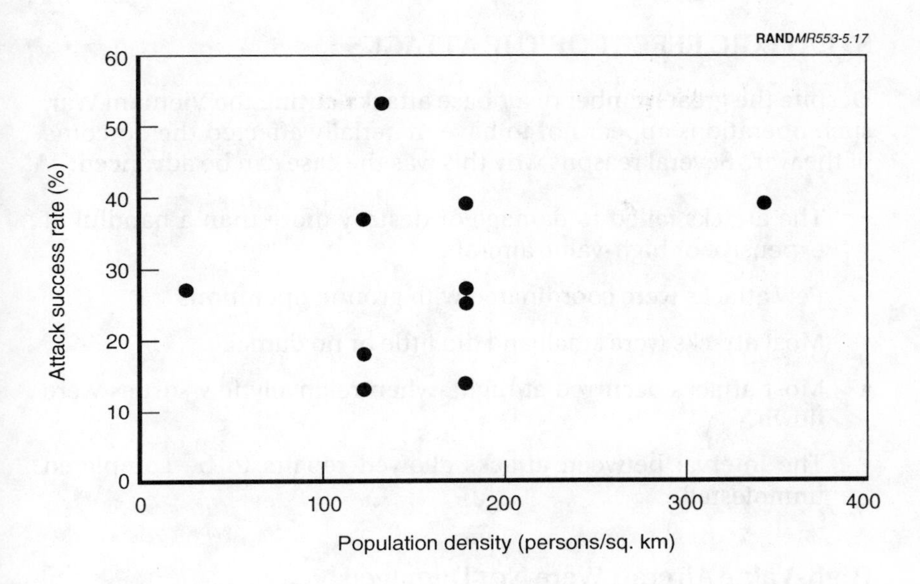

Figure 5.17—Vietnamese Provincial Population Densities Versus Attack Success Rates

tacks. Other variables that could account for the differences in success rate are quality of Viet Cong intelligence networks, access to agents on a base, size and quality of nearby VC/NVA units, and local topography and vegetation. In contrast to the bases that were so regularly attacked, Phu Cat and Tuy Hoa were rarely attacked and received fewer than 100 incoming rounds during the entire war. Phu Cat is an interesting case; it was considered difficult to defend because of its 16-mile perimeter surrounded by mountains, rice paddies, elephant grass, and swamps. Within the perimeter were hills, trees, and heavy undergrowth.[41] Yet, for whatever reason, the Viet Cong largely ignored it.

---

[41] USAF, 1969a, pp. 8–9.

STRATEGIC EFFECT OF THE ATTACKS

Despite the great number of air base attacks during the Vietnam War, such operations appear not to have materially affected the outcome of the war. Several reasons why this was the case can be advanced:

- The attacks failed to damage or destroy more than a handful of expensive or high-value aircraft.
- Few attacks were coordinated with ground operations.
- Most attacks were small and did little or no damage.
- Most attacks occurred at night, when relatively few sorties were flown.
- The interval between attacks allowed repairs to be completed unmolested.

High-Value Aircraft Were Not Damaged

A shortage of ramp space and concern about security led the USAF to base most high-value aircraft outside the Republic of Vietnam. All B-52s, KC-135s, F-105s, F-105Gs, and AC-130s were based in Thailand or Guam; strategic airlifters flew in from the United States and usually departed within hours. Inside Vietnam, the VC/NVA had trouble destroying the better-protected aircraft, such as F-4s. Those losses that were inflicted contributed to the overall attrition problem and, in some cases, may have constrained operations. But in relation to total losses (only 4 percent of total USAF losses), they were not significant. Had NVA attacks against USAF bases in Thailand succeeded in destroying Wild Weasels, gunships, B-52s, or tankers, they could have disrupted operations against North Vietnam and the Ho Chi Minh Trail, although even those losses could have been replaced by the huge USAF aircraft inventory.

Coordination with Ground Operations Was Poor

If the VC had been able to coordinate attacks against air bases with their other ground operations, such attacks might have been more effective. Their attacks on Tan Son Nhut and Bien Hoa during the Tet offensive may have been coordinated, although it appears that these targets were chosen more for their symbolic value. Even if the VC had been able to shut down operations at an individual base for the duration of a major battle, the range, depth, and diversity of U.S. airpower in Vietnam (USAF, Army, Marine, and Navy) would have given U.S. commanders alternative means of supporting ground troops.

Attacks Were Small and Occurred Mostly at Night

More generally, the small size of most attacks resulted in little or no damage. Also, most attacks occurred at night, when, with the exception of a few specialized aircraft, air operations were reduced anyway. Typically, the damage was cleaned up and the base was fully operational by the next morning. A few attacks did hit munitions or fuel storage, causing huge explosions and fires and closing bases for hours and even days.

Interval Between Attacks Enabled Recovery

The interval between attacks also contributed to the ease of recovery. At Da Nang, the most frequently attacked USAF base in Vietnam, only 16 of the 95 attacks occurred within 48 hours of another attack. Relatively large and effective attacks would have had to be waged every day or two to substantially reduce operations at a base; doing so at a single base, let alone at multiple bases, was well beyond the resources of the VC and NVA. Another way of thinking about attack spacing is to calculate the number of *base days* in Vietnam by multiplying the number of bases by the number of days the USAF was in Vietnam. There were 10 USAF MOBs operating 365 days a year for 10 years, or 36,500 base days. Air base attacks occurred on only 500 base days. Thus, USAF bases had a total of 36,000 attack-free days on which to operate.

Air base attacks did, however, raise the cost of the war—in aircraft, materiel, lives, and dollars—for the United States and the Republic of Vietnam. Such attacks forced the United States to spend considerable funds on countermeasures, to deploy large defensive forces, and to accept the inconveniences and inefficiencies that security measures can impose on the normal functioning of an airfield.

CONCLUSIONS

Whatever history's judgment on the strategic importance of the air base attacks, MACV and USAF leaders took such attacks quite seriously. The USAF developed new tactics, equipment, facilities, and organizations to counter the threat they posed. How successful were these measures in detecting attacks and reducing base vulnerability? At the least, the widespread use of minefields, fencing, lighting, and other defensive measures raised the cost and difficulty to the enemy of air base attacks.

Penetrating Attacks

Against penetrating attacks, these measures appear to have been quite successful. The limited number of sapper attacks (only 4 percent of all attacks) and the small amount of damage they inflicted suggest that USAF perimeter and interior defenses were quite robust.

Standoff Attacks

In contrast, the standoff threat, particularly from rockets, proved troublesome through the end of the war. Given the nature of the conflict and the terrain, there was no foolproof countermeasure to this threat. Nevertheless, three additional steps could have been taken to significantly reduce losses from rocket attacks.

The first would have been to integrate air base defenses more fully into rear-area security efforts. In particular, American ground forces could have been dedicated to patrolling the rocket belt around each base. Ironically, MACV appeared determined to do so in 1965 but ultimately concluded that ground forces could not be spared for duties associated with defending static locations.[42] As discussed earlier in this chapter, both Vietnamese and U.S. ground forces contributed to air base defense, but no high-quality ground forces were dedicated to this mission.

The second step would have been to increase the pace of the successful shelter program for tactical aircraft. Many tactical aircraft

---

[42]See Fox, 1979, pp. 20–28.

were lost because of the shortage of shelters. Indeed, when the shelter program ended in 1970, only 373 shelters were available in Vietnam for the 1,164 permanently assigned USAF aircraft.[43]

Third, revetments around aircraft too large for shelters would have been a great improvement over parking the aircraft on open ramps.

---

[43]There were an additional 1,000 revetments. See Fox, 1979, pp. 68–73.

Chapter Six

CONCLUSIONS

This chapter integrates lessons learned from the case studies and presents summary statistics for all 645 attacks to offer helpful historical insights to USAF officers responsible for air-base-defense planning. It begins with more tactical issues of insertion and attack modes, defense deficiencies, and strategic effects, then moves to broader observations about future air base attacks.

ATTACK TACTICS

Insertion Mode

As Figure 6.1 shows, air base attackers have used several modes of transportation for insertion. Virtually all attacks used foot travel at some point, with the exception of some motorized raids on airfields during World War II. Indeed, all 493 attacks from the Vietnam War were conducted by forces unaided by motorized vehicles. Viet Cong and NVA forces often used bicycles and boats to transport personnel and equipment; they probably used them for air base attack preparations also, but we have no means of counting the occurrence of use.

Figure 6.1 excludes the Vietnam data so that we can get a picture of other techniques. Foot travel was the most common insertion technique in the other conflicts also, closely followed by vehicle and foot insertion, primarily from the LRDG/SAS operations in North Africa. It is also interesting that 20 percent of the non-Vietnam attacks used aircraft. All submarine insertions are from World War II. Little information is available about insertion techniques for several of the

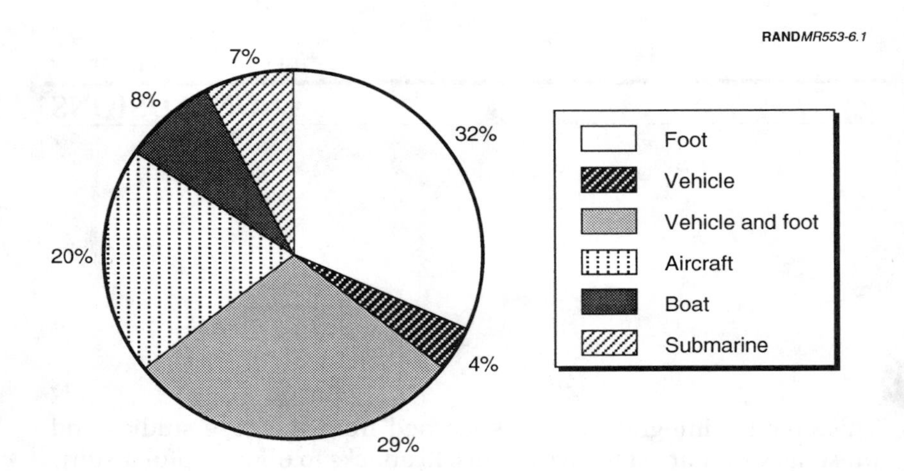

Figure 6.1—Insertion Techniques, 1940–1992 (except Vietnam)

post-Vietnam attacks listed in Table B.4. From what we know about the conflicts, topography, and forces in El Salvador, Afghanistan, the Philippines, and Iraq, it is probably safe to assume that the attacks were made on foot. Future attacks are likely to use these conventional techniques. Other means may be used as well, including commercial vehicles and more exotic options, such as high-performance parachutes, ultralight aircraft, and hang gliders.

Mode of Attack

Figure 6.2 shows the distribution of attack tactics for the 645 attacks identified in this report. Of particular interest is the apparent evolution of air base attacker tactics since World War II. Recall that all the British attacks on Axis airfields penetrated the defenses. In contrast, faced with extensive minefields, fencing, guard posts, and lights, Viet Cong and NVA attackers rarely used penetrating tactics, relying on standoff weapons for 96 percent of their attacks.

Recent attacks have used both techniques. Kurdish and Filipino insurgents used penetrating tactics; insurgents in El Salvador and Afghanistan used standoff weapons. The SAS attack against the Argentine airstrip on Pebble Island used both techniques, opening

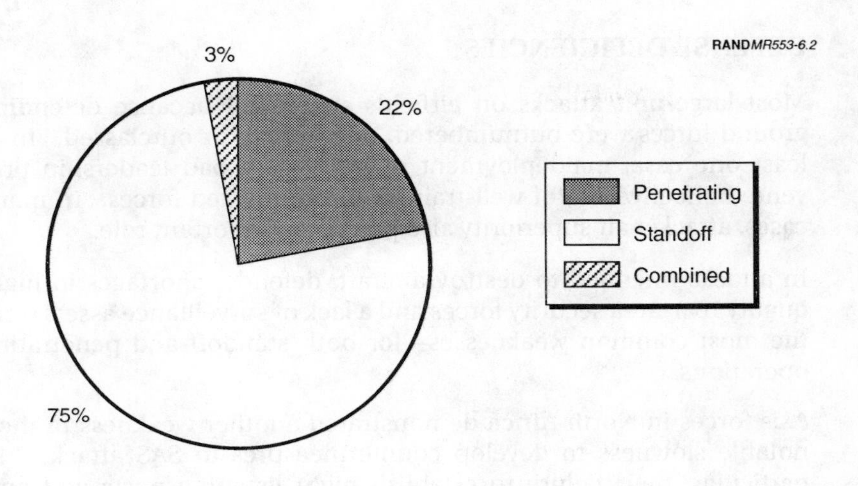

Figure 6.2—Air Base Attack Tactics, 1940–1992

the attack with naval gunfire and light antitank weapons, then moving onto the airfield to plant charges on aircraft. It is likely that both tactics will continue to be used in the future, depending on the quality of perimeter defenses. Where perimeter defenses are weak, sappers will probably continue to penetrate and place charges.

More troublesome is the possibility that precision-guided munitions (PGMs) for existing standoff weapons, as well as for some new weapons, may give small standoff attacks a lethality that they lacked in the past. For example, recall that in Vietnam almost 300 of the standoff attacks fired fewer than 10 rounds. If the attackers had been armed with PGMs for their mortars, USAF losses from these small attacks could have been very high.

As one would expect, capturing an air base or denying its use has required larger forces, typically of regimental strength. In contrast, quite small forces have been used in efforts to destroy aircraft and equipment. These attacks are typically conducted by platoons, often divided into squads or even smaller teams. The SAS used 3-to-5-man teams quite successfully in World War II; later operations appear to favor platoon- or company-size teams. Importantly, large forces are not required to conduct the most common type of air base attack; small attacks have proven to be quite effective.

DEFENSE DEFICIENCIES

Most large-unit attacks on airfields succeeded because defending ground forces were outnumbered, outgunned, or outclassed. In at least one case, maldeployment of forces and bad leadership prevented effective use of well-trained and motivated forces. In many cases, attacker air superiority also played an important role.

In attacks intended to destroy aircraft, defender shortages in high-quality rear-area security forces and a lack of surveillance assets were the most common weaknesses—for both standoff and penetrating operations.

Axis forces in North Africa demonstrated another weakness in their notable slowness to develop countermeasures to SAS attacks. In particular, their failure to establish night listening posts and ambushes outside of airfield perimeters is perplexing, since it would not have taken large forces to do so and could have paid large dividends.

Conversely, U.S. forces in Vietnam demonstrated great innovation and creativity in their defensive countermeasures. Joint-force responses to the sapper threat proved quite effective. MACV's refusal to make air base defense a high priority for resources, however, made it impossible to counter the standoff threat effectively. Without ground forces and airborne surveillance assets dedicated to controlling the standoff footprint, USAF bases remained vulnerable to the end of the war.

Reliance on other services for the defense of air bases was a problem for the RAF on Crete, the *Luftwaffe* in North Africa, and the USAF in Vietnam. In each case, air base defense had to compete with other missions on which ground commanders placed higher priority. On Crete, air base defense was also hampered by a failure to appreciate that air bases were key terrain that the attacker must be denied at all costs. In North Africa, *Luftwaffe* units reported up their own chain of command and were not integrated under General Rommel, the theater commander, which hampered the coordination of defenses.

STRATEGIC EFFECT OF THE ATTACKS

What effect have ground attacks had on the outcome of the three subject conflicts? At the very least, they caused the loss of valuable aircraft, materiel, and personnel, and forced the defenders to devote substantial resources to the defense of their airfields.

In the case of British special forces' attacks on Axis airfields in North Africa, the loss of aircraft was so severe and the airpower balance so precarious that they may have influenced the outcome of the campaign. The loss of airfields to attacking forces in other cases enabled the attacker's air force to move in and extend its range. In the Pacific theater, the need to capture and defend airfields drove both American and Japanese campaign planning. For example, the Japanese victory over the British in Malaya was made possible because ground forces had captured critical air bases. The U.S. island-hopping campaign was focused on capturing airfields; toward the end of the war, Tinian, Okinawa, and Ie Shima were captured to launch air attacks against the Japanese homeland. The Japanese attack on Midway sought to capture the island for its airfield; their failure to do so and the losses they incurred in the process marked a turning point in the war.

It is clear from this report's analysis that ground attacks on airfields in past conflicts cannot be dismissed as insignificant or rare events. The simple-but-effective tactics and the strategic rationale for the attacks are as relevant today as they were in 1940. Indeed, the centrality of airpower to modern warfare makes airfields even more tempting targets than in the past.

BACK TO THE FUTURE

The main points of this report can be summarized as follows:

- The most common air base attack objective was to destroy aircraft.
- Seventy-five percent of such attacks used standoff weapons.
- Standoff attacks have proved extremely difficult to counter.

- Reliance on non–air force services for air base defense proved problematic for the RAF on Crete, the *Luftwaffe* in North Africa, and the USAF in Vietnam.
- Small forces using unsophisticated weapons have successfully destroyed or damaged over 2,000 aircraft.

During World War II, ground attacks on air bases pursued three of the four objectives discussed in Chapter Two: capture an airfield, deny use of the airfield, and damage and destroy aircraft. During the Vietnam War, virtually all air base attacks were focused on only two objectives: destroy aircraft and harass defenders. Of the 19 attacks since Vietnam, 12 have sought to destroy aircraft. The remaining 7 captured airfields either for use as airheads or to destroy collocated ground units.

Airborne attacks since Vietnam—by the Soviet Union in Afghanistan and the United States in Grenada and Panama—may not represent future air base attacks, because few other nations have a similar capability to assault and capture, or incapacitate, an airfield. To the extent that we wish to look to historical experience as a predictor of future challenges, these cases are probably misleading: It appears highly unlikely that USAF bases will be assaulted by large airborne forces in the near future. Airborne insertion of special forces is another matter and a distinct possibility in, for instance, a future Korean conflict.

Although the possibility of large-unit attacks on airfields should not be discounted completely, it is more a prospect for adversaries of the United States than for the United States. The threat facing USAF bases in future contingencies will likely resemble those presented by SAS operations in WW II or the VC/NVA in Vietnam. If history is any indication, standoff threats will continue to pose a particularly daunting challenge. New precision-guided munitions for mortars and other standoff weapons will only exacerbate this problem.

In conclusion, attacks by small forces with the limited objective of destroying aircraft have succeeded in destroying or damaging over 2,000 aircraft between 1940 and 1992. This fact is powerful testimony to the effectiveness of small units using unsophisticated weapons against typical air base defenses and is a sobering precedent for those responsible for defending USAF bases against this threat.

Appendix A

CHURCHILL'S MEMO ON DEFENSE OF AIR BASES

June 29, 1941: Memo to Secretary of State for Air and Chief of Air Staff[1]

Further to my minute of June 20, about the responsibility of the Air Force for the local and static defence of aerodromes. Every man in Air Force uniform ought to be armed with something—a rifle, a tommy-gun, a pistol, a pike, or a mace; and every one, without exception, should do at least one hour's drill and practice every day. Every airman should have his place in the defence scheme. At least once a week an alarm should be given as an exercise (stated clearly beforehand in the signal that it is an exercise), and every man should be at his post. 90 per cent should be at their fighting stations in five minutes at the most. It must be understood by all ranks that they are expected to fight and die in the defence of their airfields. Every building which fits in with the scheme of defence should be prepared, so that each has to be conquered one by one by the enemy's parachute or glider troops. Each of these posts should have its leader appointed. In two or three hours the troops will arrive; meanwhile every post should resist and must be maintained—be it only a cottage or a mess—so that the enemy has to master each one. This is a slow and expensive process for him.

2. The enormous mass of non-combatant personnel who look after the very few heroic pilots, who alone in ordinary circumstances do all the fighting, is an inherent difficulty in the organisation of the Air

---

[1]Winston S. Churchill, *The Second World War, Volume III: The Grand Alliance*, Boston: Houghton Mifflin, 1985 ed., pp. 692–693.

Force. Here is the chance for this great mass to add a fighting quality to the necessary services they perform. Every airfield should be a stronghold of fighting air-groundmen, and not the abode of uniformed civilians in the prime of life protected by detachments of soldiers.

3. In order that I may study this matter in detail, let me have the exact field state of Northolt Aerodrome, showing every class of airman, the work he does, the weapons he has, and his part in the scheme of defence. We simply cannot afford to have the best part of half a million uniformed men, with all the prestige of the Royal Air Force attaching to them, who have not got a definite fighting value quite apart from the indispensable services they perform for the pilots.

Appendix B

CHRONOLOGY OF GROUND ATTACKS ON AIR BASES

Table B.1

Chronology of Ground Attacks on Air Bases: World War II

Event	Date	Location	Objective	Description	Sources
1	4/9/40	Aalborg, Denmark	Seize airfield as airhead	1 platoon German paratroopers (paras) supported by airlanded troops captured airfield.	Cluxton, 1967, p.x
2	4/9/40	Sola, Norway	Seize airfield as airhead	Airfield captured. Total airborne/airlanded force: 2,500 men.	Greiss, 1984b, p.29; Greiss, 1985b, Map 8a
3	4/9/40	Oslo, Norway	Seize airfield as airhead	Planned airborne assault. Ground fog and anti-aircraft artillery (AAA) drove off transports.	Cluxton, 1967, p.xi
				A few landed under fire and captured airfield; airborne assault cancelled. Total airlanded force: 3,000 men.	Greiss, 1985b, Map 8a
4	4/9/40	Stavanger, Norway	Seize airfield as airhead	1 company German paras captured airfield; airlanded troops followed.	Cluxton, 1967, p.xi; Greiss, 1984b, p.29
5	5/10/40	The Hague (1), Holland	Seize airfield as airhead	German paras captured 3 airfields. Dutch reserve forces drove them off airfields. Relieved by German ground forces 5 days later.	Cluxton, 1967, p.xvii
6	5/10/40	The Hague (2), Holland	Seize airfield as airhead	See Event 5.	See Event 5.
7	5/10/40	The Hague (3), Holland	Seize airfield as airhead	See Event 5.	See Event 5.

Table B.1—continued

Event	Date	Location	Objective	Description	Sources
8	5/10/40	Rotterdam, Holland	Seize airfield as airhead	Captured Waalhaven, Rotterdam's airport.	Cluxton, 1967, p.xvii; Fuller, 1948, p.65
9	Sept 40	Near Kufra, Libya	Destroy equipment	British Long Range Patrol (LRP) discovered auxiliary airstrips; destroyed fuel dumps and pumping equipment.	Shaw, 1989 ed., p.42
10	Oct 40	Ain Zwaya, Libya	Destroy aircraft	LRP discovered one unguarded Axis bomber (Savoia S.79) on an airstrip; set it and 160 drums of gas on fire.	Kay, 1949, p.6; Shaw, 1989 ed., p.49
11	1/11/41	Murzuk, Libya	Destroy aircraft	LRP raided airfield and fort at Murzuk. Set fire to hangar, destroying 3 Italian Ghibli bombers.	Kay, 1949, p.8; Shaw, 1989 ed., p.61
12	2/7/41	Kufra	Destroy aircraft	Free French forces and Long Range Desert Group (LRDG) attacked Italian base at Kufra, capturing airfield and fort and destroying 1 aircraft.	Shaw, 1989 ed., p.75
13	May 2–7, 1941	Habbaniya, Iraq	Deny defenders use of airfield	Pro-German Iraqi air and ground forces attacked Royal Air Force (RAF) base at Habbaniya, Iraq. RAF armored-car units and aircraft destroyed Iraq Air Force and stopped ground attack.	Churchill, 1985 ed., pp.226–233
14	5/20/41	Maleme, Crete	Seize airfield as airhead	German paras and glider troops assaulted airfield at Maleme, reinforced by airlanded troops. Captured airfield.	Beevor, 1994, pp.119–127, 144–155
15	5/20/41	Heraklion, Crete	Seize airfield as airhead	German paras assaulted airfield; driven off by defenders.	Beevor, 1994, pp.129–143, 144–155

Table B.1—continued

Event	Date	Location	Objective	Description	Sources
16	5/20/41	Retimo, Crete	Seize airfield as airhead	German paras assaulted airfield; driven off by defenders.	Beevor, 1994, pp.129–143, 144–155
17	Jun 41	Gazala, Libya	Destroy aircraft	8 Commando attempted to raid two airfields in Gazala area three times in June. Infiltrating by boat, unable to find landing place.	Messenger, 1985, pp. 109–110
18	Jun 41	Gazala, Libya	Destroy aircraft	8 Commando tried Gazala raid again; still unable to find landing place.	Messenger, 1985, p.110
19	Jun 41	Gazala, Libya	Destroy aircraft	8 Commando raiders on boats driven off by shore battery fire.	Messenger, 1985, p.110
20	11/16/41	Tmimi/Gazala, Libya	Destroy aircraft	First mission by British Special Air Service. Airborne insertion to destroy aircraft at 5 Axis airfields near Gazala. Aircraft became lost in a storm; dropped SAS far from objectives. 38 members of 62-man force killed or captured. No aircraft damaged.	Gordon, 1987, pp.81–83; Strawson, 1984, p.252; Cooper, 1991, pp.22–29
21	12/8/41	Patani, Thailand	Capture airfield for offensive ops	Airfield captured by elements of Japanese 5th division.	Kreis, 1988, p.101; Greiss, 1985a, Map 7
22	12/8/41	Singora, Thailand	Capture airfield for offensive ops	Airfield captured by elements of Japanese 5th division.	Kreis, 1988, p.101; Greiss, 1985a, Map 7

Table B.1—continued

Event	Date	Location	Objective	Description	Sources
23	12/8/41	Kota Bharu, Malaya	Capture airfield for offensive ops	RAF airfield at Kota Bharu (home for 36th RAF and 1st Royal Australian Air Force [RAAF] squadrons) captured by 56th Regiment (5,500 men) of Japanese 18th Division.	Kreis, 1988, p.101; Hart, 1977, p.225; Greiss, 1985a, Map 7
24	12/11/41	Alor Star, Malaya	Capture airfield for offensive ops	Airfield (home for RAF Squadron 62) captured by Japanese ground forces.	Kreis, 1988, p.101
25	12/13/41	El Agheila, Libya	Destroy aircraft	No aircraft on field; SAS force destroyed trucks on road.	Strawson, 1984, p. 253; Cooper, 1991, p.34
26	12/14/41	Sirte, Libya	Destroy aircraft	Aircraft were evacuated from airfield while SAS was conducting pre-raid reconnaissance. All aircraft gone by dusk.	Strawson, 1984, pp.42, 253
27	12/14/41	Tamet, Libya	Destroy aircraft	SAS destroyed 24 aircraft and fuel dump at Tamet airfield (near Sirte).	Strawson, 1984, p. 253; Shaw, 1989 ed., p.124
28	12/16/41	Sungei Patani, Malaya	Capture airfield for offensive ops	Airfield (home for RAAF & RAF Squadrons 21 &27) captured by Japanese ground forces.	Kreis, 1988, p.101; Greiss, 1985a, Map 7
29	12/18/41	Agedabia, Libya	Destroy aircraft	SAS destroyed 37 aircraft, mainly obsolete Italian CR-42 fighters, also some Stukas and Me-109s.	Cooper, 1991, p.36; Strawson, 1984, p. 253; Shaw, 1989 ed., p.126; Gordon, 1987, p.91

Table B.1—continued

Event	Date	Location	Objective	Description	Sources
30	12/23/41	Wake Island	Capture airfield for offensive ops	Japanese ground forces captured Wake Island.	Greiss, 1984a, p. 65
31	12/23/41	Sirte, Libya	Destroy aircraft	SAS found airfield defenses alerted; could not penetrate.	Strawson, 1984, p.253; Gordon, 1987, p.90
32	12/24/41	Tamet, Libya	Destroy aircraft	SAS used satchel charges to destroy 27–30 aircraft and a fuel dump. Assaulted barracks, killing ground crews.	Strawson, 1984, p.253; Gordon, 1987, p.90
33	12/24/41	Marble Arch, Libya	Destroy aircraft	SAS found no aircraft.	Strawson, 1984, p.254
34	12/26/41	Nofilia, Libya	Destroy aircraft	First satchel charge detonated prematurely; SAS forced to abort remainder of raid. 2 aircraft destroyed.	Strawson, 1984, p.253; James, 1991, p.26
35	2/14/42	Palembang, Sumatra	Capture airfield for offensive ops	Japanese paratroopers captured airfield defended by RAF Regiment.	Tucker, unpubl.
36	3/8/42	Benina, Libya	Destroy aircraft	SAS found only junked aircraft.	Strawson, 1984, p.254
37	3/8/42	Berka No. 1, Libya	Destroy aircraft	SAS team assigned to attack Berka could not find airfield.	Hoe, 1992, p.143
38	3/8/42	Berka No. 2, Libya	Destroy aircraft	SAS destroyed 15 aircraft.	Strawson, 1984, p.254; Hoe, 1992, p.143

Table B.1—continued

Event	Date	Location	Objective	Description	Sources
39	3/8/42	Barce, Libya	Destroy aircraft	SAS destroyed 1 aircraft and some trucks.	Strawson, 1984, p.254 Hoe, 1992, p.143
40	3/8/42	Slonta, Libya	Destroy aircraft	SAS unable to penetrate airfield defenses.	Strawson, 1984, p.254; Hoe, 1992, p.142
41	3/24/42	Benina, Libya	Destroy aircraft	SAS destroyed 5 aircraft in hangars; most aircraft on field were decoys.	Strawson, 1984, p.254
42	June 3–6, 1942	Midway Island	Capture airfield for offensive ops	Japanese First Air Fleet and Midway Occupation Force attacked Midway. Midway-based and carrier-based aircraft sank four Japanese carriers; stopped invasion force before any troops landed.	Greiss, 1984a, pp.111–115
43	6/9/42	Kastelli Pediados, Crete	Destroy aircraft	Special Boat Service (SBS) infiltrated Crete by submarine and raft. Used satchel charges to destroy 8 aircraft & 200 tons of aviation fuel at Kastelli Pediados.	Strawson, 1984, p.255; Beevor, 1994, p.261
44	6/12/42	Tymbaki, Crete	Destroy aircraft	SBS found Tymbaki abandoned.	Beevor, 1994, p.261; Lodwick, 1990, p.34
45	6/12/42	Maleme, Crete	Destroy aircraft	SBS found Maleme defenses included electrified fence; unable to penetrate.	Beevor, 1994, p.261; Lodwick, 1990, p.34

Table B.1—continued

Event	Date	Location	Objective	Description	Sources
46	6/13/42	Heraklion, Crete	Destroy aircraft	SAS used satchel charges to destroy or damage 21 Ju-88 bombers.	Beevor, 1994, p.262; Strawson, 1984, p.254
47	6/13/42	Benina, Libya	Destroy aircraft	SAS used satchel charges to destroy 20 Me-110 fighters, Ju-52 transports, and 30 engines found in 2 hangars. Also threw a grenade in pilot's quarters, killing at least one *Luftwaffe* officer.	Hoe, 1992, pp.164–165, 177; Strawson, 1984, p.254
48	6/13/42	Berka No. 1, Libya	Destroy aircraft	SAS destroyed 11 aircraft with satchel charges. Spotted by sentries; had to fight their way out.	Hoe, 1992, p.171
49	6/13/42	Berka No. 2, Libya	Destroy aircraft	SAS got into firefight; retreated. Destroyed 1 aircraft.	Shaw, 1989 ed., p.166; Gordon, 1987, p.105
50	6/13/42	Derna, Libya	Destroy aircraft	German Jews in British Special Interrogation Group were assigned to attack German aircraft at Derna. Got into firefight before reaching objective; all but two killed.	Gordon, 1987, p.106; Strawson, 1984, p.255
51	6/13/42	Barce, Libya	Destroy aircraft	SAS blew up fuel dump.	Hoe, 1992, pp.167–168
52	7/7/42	Bagush, Egypt	Destroy aircraft	Primers failed on half of satchel charges. SAS drove jeeps down flightline and shot up most remaining aircraft. 22 aircraft destroyed by satchel charges, 15 by machine gun fire (37 total).	Strawson, 1984, pp.69, 256; Cooper, 1991, pp. 58–62

Table B.1—continued

Event	Date	Location	Objective	Description	Sources
53	7/7/42	Fuka, Egypt	Destroy aircraft	British-French SAS teams destroyed 15 aircraft.	Strawson, 1984, p.256; James, 1991, pp.115–116
54	7/7/42	Sidi Barrani, Egypt	Destroy aircraft	SAS turned away by airfield defenses.	Strawson, 1984, p.256
55	7/7/42	El Daba, Egypt	Destroy aircraft	British and French SAS turned away by defenses.	Strawson, 1984, p.256
56	7/12/42	Fuka No. 1, Egypt	Destroy aircraft	SAS team led by Mayne. 22 aircraft destroyed at Fuka Nos. 1 & 2.	Strawson, 1984, p.256
57	7/12/42	Fuka No. 2, Egypt	Destroy aircraft	SAS team led by Jordan destroyed some aircraft. See Event 56.	Strawson, 1984, p.256
58	7/12/42	Fuka No. 3, Egypt	Destroy aircraft	SAS team led by Fraser unable to penetrate defenses.	Strawson, 1984, p.256
59	7/12/42	El Daba No. 1, Egypt	Destroy aircraft	SAS led by Jellicoe turned away by airfield defenses.	Strawson, 1984, p.256
60	7/12/42	El Daba No. 2, Egypt	Destroy aircraft	SAS led by Martin turned away by airfield defenses.	Strawson, 1984, p.256
61	7/26/42	Sidi Haneish, Egypt	Destroy aircraft	50 British and French SAS in 18 armed jeeps drove down flightline of Landing Ground 126 at night, firing incendiary ammunition into parked aircraft, destroying 40.	Cooper, 1991, pp.62–65; James, 1991, pp.154–164; Strawson, 1984, pp.65–67, 256; Hoe, 1992, pp.184–188

Table B.1—continued

Event	Date	Location	Objective	Description	Sources
62	Early August	Bagush, Libya	Destroy aircraft	LRDG used jeep raid technique, firing machine guns into parked aircraft. Raid destroyed 15 Me-109fs.	Shaw, 1989 ed., p.179; Gordon, 1987, p.112
63	8/11/42	Unknown, Sicily	Destroy aircraft	SBS raid against Axis airfield in Sicily was detected by sentries. Entire team captured; no aircraft destroyed.	Ladd, 1983, p.31
64	Aug 42–Feb 43	Henderson Fld, Guadalcanal	Deny defender use of airfield	Japanese ground forces made many attempts to capture airfield and shelled airstrip routinely.	Greiss, 1984a, p.130
65	9/12/42	Maritza, Rhodes	Destroy aircraft	SBS team inserted by submarine destroyed approximately 10 bombers used to attack Allied convoys.	Ladd, 1983, pp.32–35; Lodwick, 1990, pp.40–44
66	9/12/42	Calato, Rhodes	Destroy aircraft	SBS teams destroyed about 10 aircraft and a fuel dump.	Ladd, 1983, pp.32–35; Lodwick, 1990, pp.40–44
67	9/13/42	Barce, Libya	Destroy aircraft	LRDG shot their way onto airfield. Destroyed some hangars & fuel trucks; destroyed or damaged 32 SM-79, Stukas, and Macchi aircraft.	Gordon, 1987, p.129; Peniakoff, 1950, p.149; Strawson, 1984, p.257
68	Oct/Nov 42	Unknown Egypt & Libya	Destroy aircraft	SAS conducted many small raids in Oct and Nov 42 against German lines of communication, supplies, and airfields with "some success."	Strawson, 1984, p.256

Table B.1—continued

Event	Date	Location	Objective	Description	Sources
69	11/8/42	Tafaraoui, Algeria	Capture airfield for offensive ops	U.S. ground forces captured Tafaraoui airfield in Algeria after short fight with Vichy French defenders.	Craven & Cate, 1949, p.72
70	11/8/42	Blida, Algeria	Capture airfield for offensive ops	British 11th Infantry Brigade captured Blida airfield near Algiers.	Strawson, 1969, p.183
71	11/8/42	Maison Blanche, Algeria	Capture airfield for offensive ops	U.S. 39th Regimental Combat Team (RCT) captured Maison Blanche airfield near Algiers.	Strawson, 1969, p.183
72	11/8/42	Hussein Day, Algeria	Capture airfield for offensive ops	U.S. 38th RCT captured Hussein Day airfield near Algiers.	Strawson, 1969, p.183
73	11/10/42	La Senia, Algeria	Deny defender use of airfield	U.S. armored forces captured Vichy French airfield at La Senia.	Craven & Cate, 1949, pp. 56, 68, 70, 73
74	11/10/42	Port Lyautey, French Morocco	Capture airfield for offensive ops	U.S. Rangers captured airfield.	Craven & Cate, 1949, p.77
75	11/12/42	Duzerville, Algeria	Deny defender use of airfield	British 3rd Parachute Battalion (312 men) conducted airborne assault of airfield near Bone. Vichy forces did not oppose.	Craven & Cate, 1949, p.79
76	11/14/42	Youks-les-Bains, Algeria	Deny defender use of airfield	U.S. 2nd Bat/503rd Parachute Regiment (350 men) conducted airborne assault. Unopposed by French.	Craven & Cate, Vol. II, pp.79–81
77	11/16/42	Souk-el-Arba, Algeria	Capture airfield for offensive ops	384 British paratroopers jumped onto airfield. No opposition.	Craven & Cate, 1949, p.81

Table B.1—continued

Event	Date	Location	Objective	Description	Sources
78	Dec 42	Souk-el-Arba, Algeria	Destroy aircraft	German special forces parachuted into Souk-el-Arba and Souk-el-Ahras areas to attack airfields and key bridges. Landed miles from objectives; British patrols captured all parachutists before they reached objectives.	Lucas, 1985, p.93
79	Jan 43	Hon, Libya	Destroy aircraft	LRDG assigned to destroy Axis aircraft at Hon and Sebha in western Libya. Heavy rain made approach route impassable.	Kay, 1950, p.11
80	Jan 43	Sebha, Libya	Destroy aircraft	Raid aborted due to weather. See Event 79.	Kay, 1950, p.11
81	Jun 43	Kastelli Pediados, Crete	Destroy aircraft	SBS attacked German airfields on Crete prior to invasion of Sicily. Destroyed 5 Stukas and Ju-88 bombers with satchel charges.	Lodwick, 1990, p.63; Beevor, 1994, p.285
82	Jun 43	Tymbaki, Crete	Destroy aircraft	SBS found no aircraft at Tymbaki (same as June 42 raid, Event 44).	Warner, 1971, p.95; Beevor, 1994, p.285
83	Jun 43	Heraklion, Crete	Destroy aircraft	SBS found no aircraft at Heraklion but did blow up a large fuel dump at Peza.	Warner, 1971, p.95
84	7/9/43	Ottana, Sardinia	Destroy aircraft	SBS team parachuted into Sardinia to raid German airfield at Ottana. Blew up fuel tanks, ammunition, and about 5 aircraft.	Strawson, 1969, p.116; Lodwick, 1990, pp.71–73; Verney, 1955, pp.174–178

Table B.1—continued

Event	Date	Location	Objective	Description	Sources
85	9/14/43	Antimachia, Island of Cos	Capture airfield for offensive ops	British paras captured Italian airfield on Island of Cos in eastern Aegean Sea.	Churchill, Vol. V, pp.184–185; Tucker, unpubl.
86	10/3/43	Antimachia, Island of Cos	Deny defender use of airfields	Over 2,000 Germans paratroopers and seaborne army troops recaptured airfield and island defended by 229 men of the RAF Regiment. Defenders armed only with rifles and submachine guns against Germans equipped with mortars and self-propelled guns, and supported by air.	Churchill, Vol. V, pp.184–185; Tucker, unpubl.
87	1/12/44	San Egidio, Italy	Destroy aircraft	2nd SAS in *Operation Pomegranate* inserted by parachute and raid German airfield. Unknown results.	Strawson, 1984, p.262
88	7/24/44	Tinian, Marianas	Capture airfield for offensive ops	U.S. ground forces captured Tinian for its airfield complex.	Craven & Cate, 1953, p.31; Greiss, 1984a, p.166
89	9/4/44	Ling Ling, China	Deny defender use of airfields	Japanese launched *Operation Ichigo* to capture U.S. 14th Air Force base in East China. Ling Ling was first to fall.	Craven & Cate, 1953, pp.220–225
90	9/23/44	Araxos, Greece	Capture airfield for offensive ops	British paratroopers and seaborne force, including No. 2908 Squadron RAF Regiment, captured Axis airfield at Araxos.	Saunders, 1954, p.243
91	9/26/44	Tanchuk, China	Deny defender use of airfields	Japanese captured U.S. airfield at Tanchuk.	See Event 89.

Table B.1—continued

Event	Date	Location	Objective	Description	Sources
92	10/10/44	Megara, Greece	Capture airfield for offensive ops	British ground forces captured Megara Airfield (20 miles west of Athens).	Saunders, 1954, p.243
93	11/10/44	Kweilin, China	Deny defender use of airfields	Japanese captured U.S. airfield at Kweilin.	See Event 89.
94	11/11/44	Liuchow, China	Deny defender use of airfields	Japanese captured U.S. airfield at Liuchow.	See Event 89.
95	Feb 19–Mar 26, 1945	Iwo Jima	Capture airfield for offensive ops	U.S. Marines captured island for airfield. USAF used existing Japanese airfield and built additional ones to support B-29 operations against the Japanese home islands.	Craven & Cate, 1953, pp.577–578
96	Mar 45	Meiktila, Burma	Capture airfield for offensive ops	British 14th Army captured Meiktila airfield from Japanese.	Oliver, 1970, p.26
97–127	Mar 45	Meiktila, Burma	Deny defender use of airfields	During the month of March, Japanese made nightly attacks on Meiktila airfield. RAF Regiment (Wing 1307) successfully defended airfield.	Oliver, 1970, p.26; Tucker, unpubl.
128	4/1/45	Kadena , Okinawa	Capture airfield for offensive ops	U.S. Marines and Army units captured Kadena airfield for USAF operations.	Greiss, 1984a, p.240
129	4/1/45	Yontan, Okinawa	Capture airfield for offensive ops	U.S. ground forces captured Yontan airfield.	Greiss, 1984a, p.240
130	4/21/45	Ie Shima	Capture airfield for offensive ops	In five days of fighting, U.S. ground forces captured Japanese airfield on Ie Shima Island (near Okinawa).	Greiss, 1984a, pp.237, 241

Table B.2

Chronology of Ground Attacks on Air Bases: Korean War

Event	Date	Location	Objective	Description	Sources
1	Aug 50	Pohang, ROK	Destroy aircraft	Ground crews successfully defended airfield perimeter at night against infiltration attempts by North Korean guerrillas. Airfield was evacuated on 13 August, one day after North Korean forces captured port of Pohang.	Futrell, 1983, p.124
2	Sept 50–Nov 50	Kunsan, ROK	Deny defender use of airfield	Following UN landing at Inchon in Sept 50, USAF attempted to open sod strip at Kunsan for air operations. Harassment by guerrillas prevented this opening for several months.	Schuetta, 1964, p.38
3	Nov 50	Multiple locations, ROK	Destroy aircraft	Guerrillas fired small arms at UN aircraft taking off and landing. No reports of damage or loss of aircraft.	Schuetta, 1964, p.38

NOTE: ROK = Republic of Korea.

Table B.3

Chronology of Ground Attacks on Air Bases: Vietnam War

Event	Date	Location	Objective	Description (aircraft [ac] losses destroyed/damaged)	Sources
1	11/1/64	Bien Hoa	Destroy aircraft	Standoff attack (70 rounds [rds]); ac losses 7/18.	Fox, 1979, p.173
2	2/7/65	Army airstrip, Pleiku	Destroy aircraft	Standoff attack with 81-mm mortars destroyed 5 helicopters.	Berger, 1984, p.37
3	2/7/65	Tuy Hoa	Destroy aircraft	Aviation fuel storage set on fire by sapper or standoff attack.	Berger, 1984, p.37
4	2/8/65	Soc Trang	Destroy aircraft	Attack on airfield caused no damage.	Berger, 1984, p.37
5	5/18/65	Qui Nhon	Destroy aircraft	Standoff attack destroyed 2 A-1 aircraft.	Berger, 1984, pp.34, 62–63
6	7/1/65	Da Nang	Destroy aircraft	Standoff/sapper attack (6 rds); ac losses 6/3.	Fox, 1979, p.173
7	8/2/65	Nha Trang	Destroy aircraft	Standoff attack (7 rds); ac losses 0/0.	Fox, 1979, p.173
8	8/23/65	Bien Hoa	Destroy aircraft	Standoff attack (97 rds); ac losses 0/11.	Fox, 1979, p.173
9	1/25/66	Da Nang	Destroy aircraft	Standoff attack (20 rds); ac losses 0/0.	Fox, 1979, p.173
10	2/17/66	Na Khang, Laos	Deny defender use of airfield	Laotian communist forces captured key Royal Laotian airfield.	Berger, 1984, p.123
11	2/20/66	Binh Thuy	Destroy aircraft	Standoff attack (5 rds); ac losses 0/1.	Fox, 1979, p.173
12	4/13/66	Tan Son Nhut	Destroy aircraft	Standoff attack (243 rds); ac losses 2/62.	USAF, 1966

Table B.3—continued

Event	Date	Location	Objective	Description (aircraft [ac] losses destroyed/damaged)	Sources
13	4/22/66	Pleiku	Destroy aircraft	Standoff attack (79 rds); ac losses 2/11.	Fox, 1979, p.173
14	7/8/66	Binh Thuy	Destroy aircraft	Standoff attack (40 rds); ac losses 1/4.	Fox, 1979, p.173
15	10/18/66	Bien Hoa	Destroy aircraft	Sapper attack; ac losses 0/0.	Fox, 1979, p.173
16	12/4/66	Tan Son Nhut	Destroy aircraft	Standoff and sapper attack (33 rds); ac losses 0/20.	Fox, 1979, p.173
17	12/24/66	Binh Thuy	Destroy aircraft	Standoff attack (29 rds); ac losses 0/2.	Fox, 1979, p.173
18	1/7/67	Pleiku	Destroy aircraft	Standoff attack (32 rds); ac losses 0/0.	Fox, 1979, p.174
19	1/12/67	Binh Thuy	Destroy aircraft	Standoff attack (67 rds); ac losses 0/14 .	Fox, 1979, p.174
20	2/2/67	Luang Prabang, Laos	Destroy aircraft	Laotian communist forces mortared Royal Laotian Air Force base, destroying 8 and damaging 3 aircraft.	Berger, 1984, p.123
21	2/7/67	Bien Hoa	Destroy aircraft	Sabotage destroyed 2,600 napalm bombs worth $342,000.	Fox, 1979, p.174
22	2/8/67	Binh Thuy	Destroy aircraft	Standoff attack (56 rds); ac losses 0/11.	Fox, 1979, p.174
23	2/15/67	Nha Trang	Destroy aircraft	Sapper attack; ac losses 3/5.	Fox, 1979, p.174
24	2/27/67	Da Nang	Destroy aircraft	Standoff attack (56 rds); ac losses 0/13.	Fox, 1979, p.174
25	3/15/67	Da Nang	Destroy aircraft	Standoff attack (unknown rds); ac losses 0/7.	Fox, 1979, p.174
26	3/27/67	Binh Thuy	Destroy aircraft	Standoff attack (35 rds); ac losses 0/2.	Fox, 1979, p.174
27	5/7/67	Binh Thuy	Destroy aircraft	Standoff attack (69 rds); ac losses 0/4.	Fox, 1979, p.174
28	5/12/67	Bien Hoa	Destroy aircraft	Standoff attack (189 rds); ac losses 4/32.	Fox, 1979, p.174
29	7/15/67	Da Nang	Destroy aircraft	Standoff attack (83 rds); ac losses 10/50.	Fox, 1979, p.174

Table B.3—continued

Event	Date	Location	Objective	Description (aircraft [ac] losses destroyed/damaged)	Sources
30	7/16/67	Luang Prabang, Laos	Destroy aircraft	Laotian communist attack destroyed 10 Royal Laotian Air Force T-28s.	Berger, 1984, p.123
31	9/2/67	Da Nang	Destroy aircraft	Standoff attack (9 rds); ac losses 0/6.	Fox, 1979, p.174
32	9/7/67	Tuy Hoa	Destroy aircraft	Automatic weapons fired; ac losses 0/0.	Fox, 1979, p.174
33	9/9/67	Da Nang	Destroy aircraft	Standoff attack (3 rds); ac losses 0/2.	Fox, 1979, p.175
34	10/10/67	Nha Trang	Destroy aircraft	Standoff attack (16 rds); ac losses 0/0.	Fox, 1979, p.175
35	11/5/67	Bien Hoa	Destroy aircraft	Standoff attack (15 rds); ac losses 0/0.	Fox, 1979, p.175
36	11/15/67	Dak To	Destroy aircraft	Rocket attack destroyed 2 C-130 transports.	Berger, 1984, p.175
37	11/26/67	Nha Trang	Destroy aircraft	Standoff attack (30 rds); ac losses 1/3.	Fox, 1979, p.175
38	Jan–Apr 68	Khe Sanh	Deny defender use of airfield	Standoff attacks destroyed 3 C-123s and 1 C-130.	Berger, 1984, p.176
39	1/3/68	Da Nang	Destroy aircraft	Standoff attack (49 rds); ac losses 1/20.	Fox, 1979, p.175
40	1/20/68	Pleiku	Destroy aircraft	Standoff attack (8 rds); ac losses 0/0.	Fox, 1979, p.175
41	1/30/68	Pleiku	Destroy aircraft	Standoff attack (13 rds); ac losses 0/2.	Fox, 1979, p.175
42	1/30/68	Da Nang	Destroy aircraft	Standoff attack (40 rds); ac losses 5/25.	Fox, 1979, p.175
43	1/31/68	Bien Hoa	Destroy aircraft	Battalion-plus assault included standoff (45 rds); ac losses 2/17.	Fox, 1979, p.175
44	1/31/68	Tan Son Nhut	Destroy aircraft	Multiple battalions assault Tan Son Nhut; ac losses 0/13.	Fox, 1979, p.175
45	1/31/68	Nha Trang	Harass	Standoff attack (2 rds); ac losses 0/0.	Fox, 1979, p.175
46	2/3/68	Binh Thuy	Destroy aircraft	Standoff attack (9 rds); ac losses 0/0.	Fox, 1979, p.175

Table B.3—continued

Event	Date	Location	Objective	Description (aircraft [ac] losses destroyed/damaged)	Sources
47	2/4/68	Binh Thuy	Destroy aircraft	Standoff attack (73 rds); ac losses 0/16.	Fox, 1979, p.175
48	2/5/68	Binh Thuy	Destroy aircraft	Standoff attack (45 rds); ac losses 0/12.	Fox, 1979, p.176
49	2/6/68	Phu Cat	Destroy aircraft	Standoff attack (10 rds); ac losses 0/0.	Fox, 1979, p.176
50	2/7/68	Binh Thuy	Destroy aircraft	Standoff attack (9 rds); ac losses 0/1.	Fox, 1979, p.176
51	2/11/68	Bien Hoa	Destroy aircraft	Standoff attack (16 rds); ac losses 6/26.	Fox, 1979, p.176
52	2/12/68	Binh Thuy	Destroy aircraft	Standoff attack (9 rds); ac losses 1/6.	Fox, 1979, p.176
53	2/13/68	Binh Thuy	Destroy aircraft	Sapper attack; ac losses 0/0.	Fox, 1979, p.176
54	2/13/68	Binh Thuy	Destroy aircraft	Standoff attack (44 rds); ac losses 0/19.	Fox, 1979, p.176
55	2/13/68	Binh Thuy	Destroy aircraft	Standoff attack (26 rds); ac losses 0/0.	Fox, 1979, p.176
56	2/16/68	Binh Thuy	Destroy aircraft	Standoff attack (26 rds); ac losses 0/0.	Fox, 1979, p.176
57	2/16/68	Nha Trang	Destroy aircraft	Standoff attack (21 rds); ac losses 0/0.	Fox, 1979, p.176
58	2/18/68	Tan Son Nhut	Destroy aircraft	Standoff attack (60 rds). See Event 82 for results.	Fox, 1979, p.176
59	2/18/68	Binh Thuy	Destroy aircraft	Standoff attack (12 rds); ac losses 0/0.	Fox, 1979, p.176
60	2/18/68	Bien Hoa	Destroy aircraft	Standoff attack (7 rds); ac losses 1/3.	Fox, 1979, p.176
61	2/18/68	Tan Son Nhut	Destroy aircraft	Standoff attack (2 rds). See Event 82.	Fox, 1979, p.176
62	2/18/68	Tan Son Nhut	Destroy aircraft	Standoff attack (2 rds). See Event 82.	Fox, 1979, p.176
63	2/18/68	Tan Son Nhut	Destroy aircraft	Standoff attack (1 rd). See Event 82.	Fox, 1979, p.176
64	2/19/68	Tan Son Nhut	Destroy aircraft	Standoff attack (2 rds). See Event 82.	Fox, 1979, p.177
65	2/19/68	Tan Son Nhut	Destroy aircraft	Standoff attack (2 rds). See Event 82.	Fox, 1979, p.177
66	2/19/68	Tan Son Nhut	Destroy aircraft	Standoff attack (3 rds). See Event 82.	Fox, 1979, p.177
67	2/19/68	Tan Son Nhut	Destroy aircraft	Standoff attack (5 rds). See Event 82.	Fox, 1979, p.177

Table B.3—continued

Event	Date	Location	Objective	Description (aircraft [ac] losses destroyed/damaged)	Sources
68	2/20/68	Tan Son Nhut	Destroy aircraft	Standoff attack (1 rd). See Event 82.	Fox, 1979, p.177
69	2/20/68	Tan Son Nhut	Destroy aircraft	Standoff attack (2 rds). See Event 82.	Fox, 1979, p.177
70	2/21/68	Tan Son Nhut	Destroy aircraft	Standoff attack (1 rd). See Event 82.	Fox, 1979, p.177
71	2/21/68	Tan Son Nhut	Destroy aircraft	Standoff attack (3 rds). See Event 82.	Fox, 1979, p.177
72	2/22/68	Pleiku	Destroy aircraft	Standoff attack (18 rds); ac losses 0/0.	Fox, 1979, p.177
73	2/23/68	Binh Thuy	Destroy aircraft	Standoff attack (56 rds); ac losses 0/3.	Fox, 1979, p.177
74	2/24/68	Tan Son Nhut	Destroy aircraft	Standoff attack (20 rds). See Event 82.	Fox, 1979, p.177
75	2/24/68	Da Nang	Destroy aircraft	Standoff attack (10 rds); ac losses 0/5.	Fox, 1979, p.177
76	2/26/68	Binh Thuy	Destroy aircraft	Standoff attack (33 rds); ac losses 1/4.	Fox, 1979, p.177
77	2/27/68	Tan Son Nhut	Destroy aircraft	Standoff attack (3 rds). See Event 82.	Fox, 1979, p.177
78	2/27/68	Tan Son Nhut	Destroy aircraft	Standoff attack (4 rds). See Event 82.	Fox, 1979, p.177
79	2/28/68	Bien Hoa	Destroy aircraft	Standoff attack (32 rds); ac losses 0/5.	Fox, 1979, p.177
80	2/28/68	Tan Son Nhut	Destroy aircraft	Standoff attack (2 rds). See Event 82.	Fox, 1979, p.178
81	3/1/68	Pleiku	Destroy aircraft	Standoff attack (11 rds); ac losses 0/0.	Fox, 1979, p.178
82	3/1/68	Tan Son Nhut	Destroy aircraft	Standoff attack (16 rds); ac losses 4/74.	Fox, 1979, p.178
83	3/4/68	Cam Ranh Bay	Destroy aircraft	Standoff attack (27 rds); ac losses 0/0.	Fox, 1979, p.178
84	3/5/68	Binh Thuy	Destroy aircraft	Standoff attack (110 rds); ac losses 2/7.	Fox, 1979, p.178
85	3/6/68	Tuy Hoa	Harass	Standoff attack (4 rds); ac losses 0/0.	Fox, 1979, p.178
86	3/6/68	Pleiku	Harass	Standoff attack (1 rd); ac losses 0/0.	Fox, 1979, p.178
87	3/7/68	Phan Rang	Destroy aircraft	Standoff attack (10 rds); ac losses 0/0.	Fox, 1979, p.178
88	3/10/68	Pleiku	Destroy aircraft	Standoff attack (7 rds); ac losses 0/3.	Fox, 1979, p.178

Table B.3—continued

Event	Date	Location	Objective	Description (aircraft [ac] losses destroyed/damaged)	Sources
89	3/12/68	Bien Hoa	Destroy aircraft	Standoff attack (7 rds); ac losses 0/0.	Fox, 1979, p.178
90	3/14/68	Binh Thuy	Destroy aircraft	Standoff attack (29 rds). Results combined with those of Event 96.	Fox, 1979, p.178
91	3/14/68	Binh Thuy	Destroy aircraft	Standoff attack (25 rds). Results combined with those of Event 96.	Fox, 1979, p.178
92	3/17/68	Binh Thuy	Destroy aircraft	Standoff attack (65 rds). Results combined with those of Event 96.	Fox, 1979, p.178
93	3/21/68	Tan Son Nhut	Destroy aircraft	Standoff attack (10 rds); ac losses 0/7.	Fox, 1979, p.178
94	3/22/68	Binh Thuy	Destroy aircraft	Standoff attack (36 rds). Results combined with those of Event 96.	Fox, 1979, p.178
95	3/22/68	Bien Hoa	Destroy aircraft	Standoff attack (9 rounds); ac losses 0/5.	Fox, 1979, p.179
96	3.25/68	Binh Thuy	Destroy aircraft	Standoff attack (85 rds); ac losses 3/29.	Fox, 1979, p.179
97	4/1/68	Tuy Hoa	Destroy aircraft	Sapper attack; ac losses 0/0.	Fox, 1979, p.179
98	4/2/68	Pleiku	Destroy aircraft	Standoff attack (21 rds); ac losses 0/0.	Fox, 1979, p.179
99	4/5/68	Bien Hoa	Destroy aircraft	Standoff attack (12 rds); ac losses 0/0.	Fox, 1979, p.179
100	4/9/68	Binh Thuy	Destroy aircraft	Standoff attack (30 rds); ac losses 0/0.	Fox, 1979, p.179
101	4/13/68	Binh Thuy	Destroy aircraft	Standoff attack (35 rds); ac losses 0/0.	Fox, 1979, p.179
102	5/3/68	Tuy Hoa	Destroy aircraft	Standoff attack (24 rds); ac losses 0/0.	Fox, 1979, p.179
103	5/5/68	Pleiku	Destroy aircraft	Standoff attack (11 rds); ac losses 2/0.	Fox, 1979, p.179
104	5/5/68	Da Nang	Harass	Standoff attack (1 rd); ac losses 0/0.	Fox, 1979, p.179
105	5/5/68	Bien Hoa	Destroy aircraft	Standoff attack (74 rds); ac losses 0/13.	Fox, 1979, p.179
106	5/5/68	Bien Hoa	Destroy aircraft	Standoff attack (7 rds); ac losses 0/0.	Fox, 1979, p.179

Table B.3—continued

Event	Date	Location	Objective	Description (aircraft [ac] losses destroyed/damaged)	Sources
107	5/6/68	Tan Son Nhut	Destroy aircraft	Standoff attack (10 rds); ac losses 0/0.	Fox, 1979, p.179
108	5/7/68	Tan Son Nhut	Destroy aircraft	Standoff attack (11 rds); ac losses 0/1.	Fox, 1979, p.179
109	5/7/68	Bien Hoa	Harass	Standoff attack (1 rd); ac losses 0/0.	Fox, 1979, p.179
110	5/8/68	Tan Son Nhut	Destroy aircraft	Standoff attack (14 rds); ac losses 0/0.	Fox, 1979, p.179
111	5/8/68	Pleiku	Destroy aircraft	Standoff attack (6 rds); ac losses 0/2.	Fox, 1979, p.179
112	5/9/68	Da Nang	Destroy aircraft	Standoff attack (4 rds); ac losses 0/1.	Fox, 1979, p.180
113	5/9/68	Da Nang	Harass	Standoff attack (3 rds); ac losses 0/0.	Fox, 1979, p.180
114	5/10/68	Tan Son Nhut	Destroy aircraft	Standoff attack (7 rds); ac losses 0/0.	Fox, 1979, p.180
115	5/11/68	Da Nang	Destroy aircraft	Standoff attack (8 rds); ac losses 0/6.	Fox, 1979, p.180
116	5/12/68	Da Nang	Destroy aircraft	Standoff attack (3 rds); ac losses 0/2.	Fox, 1979, p.180
117	5/21/68	Binh Thuy	Destroy aircraft	Standoff attack (40 rds); ac losses 1/0.	Fox, 1979, p.180
118	5/22/68	Nha Trang	Destroy aircraft	Standoff attack (5 rds); ac losses 0/4.	Fox, 1979, p.180
119	5/23/68	Binh Thuy	Harass	Standoff attack (3 rds); ac losses 0/0.	Fox, 1979, p.180
120	5/24/68	Binh Thuy	Destroy aircraft	Standoff attack (40 rds); ac losses 0/0.	Fox, 1979, p.180
121	5/29/68	Da Nang	Destroy aircraft	Standoff attack (8 rds); ac losses 0/8.	Fox, 1979, p.180
122	6/12/68	Tan Son Nhut	Destroy aircraft	Standoff attack (13 rds); ac losses 2/12	Fox, 1979, p.180
123	6/14/68	Tan Son Nhut	Destroy aircraft	Standoff attack (4 rds); ac losses 0/2.	Fox, 1979, p.180
124	6/15/68	Bien Hoa	Destroy aircraft	Standoff attack (9 rds); ac losses 0/6.	Fox, 1979, p.180
125	6/21/68	Nha Trang	Destroy aircraft	Standoff attack (11 rds); ac losses 0/4.	Fox, 1979, p.180
126	6/23/68	Phan Rang	Destroy aircraft	Standoff attack (18 rds); ac losses 0/5.	Fox, 1979, p.180
127	6/24/68	Binh Thuy	Destroy aircraft	Standoff attack fires (10 rds); ac losses 0/0.	Fox, 1979, p.180

Table B.3—continued

Event	Date	Location	Objective	Description (aircraft [ac] losses destroyed/damaged)	Sources
128	6/26/68	Binh Thuy	Destroy aircraft	Standoff attack (35 rds); ac losses 0/0.	Fox, 1979, p.181
129	7/23/68	Da Nang	Destroy aircraft	Standoff attack (16 rds); ac losses 0/7.	Fox, 1979, p.181
130	7/23/68	Da Nang	Harass	Standoff attack (1 rd); ac losses 0/0.	Fox, 1979, p.181
131	7/26/68	Udorn AFB, Thailand	Destroy aircraft	Sapper attack caused heavy damage to a C-141, moderate damage to an F-4D, and light damage to an HH-43.	USAF, 1968
132	7/27/68	Da Nang	Destroy aircraft	Standoff attack (6 rds); ac losses 1/4.	Fox, 1979, p.181
133	7/29/68	Tuy Hoa	Destroy aircraft	Sapper attack; ac losses 2/7.	Fox, 1979, p.181
134	8/21/68	Phan Rang	Destroy aircraft	Standoff attack (27 rds); ac losses 0/2.	Fox, 1979, p.181
135	8/22/68	Binh Thuy	Destroy aircraft	Standoff attack (35 rds); ac losses 0/0.	Fox, 1979, p.181
136	8/22/68	Binh Thuy	Destroy aircraft	Standoff attack (22 rds); ac losses 0/0.	Fox, 1979, p.181
137	8/22/68	Bien Hoa	Destroy aircraft	Standoff attack (11 rds); ac losses 0/0.	Fox, 1979, p.181
138	8/23/68	Pleiku	Destroy aircraft	Standoff attack (17 rds); ac losses 0/4.	Fox, 1979, p.181
139	8/23/68	Da Nang	Destroy aircraft	Standoff attack (13 rds); ac losses 0/3.	Fox, 1979, p.181
140	8/24/68	Binh Thuy	Destroy aircraft	Standoff attack (12 rds); ac losses 0/0.	Fox, 1979, p.181
141	8/25/68	Binh Thuy	Destroy aircraft	Standoff attack (33 rds); ac losses 0/0.	Fox, 1979, p.181
142	8/25/68	Binh Thuy	Destroy aircraft	Standoff attack (29 rds); ac losses 0/0.	Fox, 1979, p.181
143	8/27/68	Da Nang	Destroy aircraft	Standoff attack (6 rds); ac losses 0/5.	Fox, 1979, p.181
144	8/29/68	Binh Thuy	Destroy aircraft	Standoff attack (44 rds); ac losses 2/38.	Fox, 1979, p.181
145	8/30/68	Bien Hoa	Harass	Standoff attack (2 rds); ac losses 0/0.	Fox, 1979, p.181
146	8/31/68	Da Nang	Destroy aircraft	Standoff attack (1 rd); ac losses 0/6.	Fox, 1979, p.182

Table B.3—continued

Event	Date	Location	Objective	Description (aircraft [ac] losses destroyed/damaged)	Sources
147	9/2/68	Da Nang	Harass	Standoff attack (1 rd); ac losses 0/0.	Fox, 1979, p.182
148	9/4/68	Da Nang	Harass	Standoff attack (2 rds); ac losses 0/0.	Fox, 1979, p.182
149	9/8/68	Bien Hoa	Destroy aircraft	Standoff attack (8 rds); ac losses 0/6.	Fox, 1979, p.182
150	9/11/68	Binh Thuy	Destroy aircraft	Standoff attack (16 rds); ac losses 0/0.	Fox, 1979, p.182
151	9/11/68	Binh Thuy	Destroy aircraft	Standoff attack (40 rds); ac losses 2/21.	Fox, 1979, p.182
152	9/18/68	Da Nang	Harass	Standoff attack (3 rds); ac losses 0/0.	Fox, 1979, p.182
153	9/21/68	Pleiku	Destroy aircraft	Standoff attack (35 rds); ac losses 0/3.	Fox, 1979, p.182
154	9/21/68	Nha Trang	Destroy aircraft	Standoff attack (23 rds). Losses included with those of Event 153.	Fox, 1979, p.182
155	9/22/68	Nha Trang	Destroy aircraft	Standoff attack (4 rds); ac losses 0/4.	Fox, 1979, p.182
156	9/29/68	Da Nang	Destroy aircraft	Standoff attack (4 rds); ac losses 0/5.	Fox, 1979, p.182
157	9/29/68	Binh Thuy	Destroy aircraft	Standoff attack (40 rds); ac losses 0/0.	Fox, 1979, p.182
158	10/26/68	Bien Hoa	Destroy aircraft	Standoff attack (7 rds); ac losses 0/0.	Fox, 1979, p.182
159	11/21/68	Pleiku	Destroy aircraft	Standoff attack (24 rds); ac losses 0/0	Fox, 1979, p.182
160	12/23/68	Pleiku	Destroy aircraft	Standoff attack (16 rds); ac losses 0/3.	Fox, 1979, p.182
161	1/10/69	Binh Thuy	Destroy aircraft	Standoff attack (62 rds); ac losses 0/0.	Fox, 1979, p.183
162	1/10/69	Binh Thuy	Destroy aircraft	Second standoff attack 1 hour later (56 rds); ac losses 0/0.	Fox, 1979, p.183
163	1/15/69	Pleiku	Destroy aircraft	Standoff attack (17 rds); ac losses 0/0.	Fox, 1979, p.183
164	1/22/69	Da Nang	Destroy aircraft	Standoff attack (26 rds); ac losses 0/0.	Fox, 1979, p.183
165	1/26/69	Phan Rang	Destroy aircraft	Standoff/sapper attack (74 rds); ac losses 2/11.	Fox, 1979, p.183

Table B.3—continued

Event	Date	Location	Objective	Description (aircraft [ac] losses destroyed/damaged)	Sources
166	1/29/69	Binh Thuy	Destroy aircraft	Sapper attack; ac losses 0/0.	Fox, 1979, p.183
167	2/22/69	Phan Rang	Destroy aircraft	Standoff attack (86 rds); ac losses 0/20.	Fox, 1979, p.183
168	2/22/69	Phu Cat	Destroy aircraft	Sapper attack; ac losses 0/0.	Fox, 1979, p.183
169	2/23/69	Bien Hoa	Destroy aircraft	Standoff attack (39 rds); ac losses 2/8.	Fox, 1979, p.183
170	2/23/69	Binh Thuy	Destroy aircraft	Standoff attack (11 rds); ac losses 0/7.	Fox, 1979, p.183
171	2/23/69	Cam Ranh Bay	Destroy aircraft	Standoff attack (7 rds); ac losses 0/6.	Fox, 1979, p.183
172	2/23/69	Da Nang	Destroy aircraft	Standoff attack (11 rds); ac losses 0/0.	Fox, 1979, p.183
173	2/23/69	Pleiku	Destroy aircraft	Standoff attack (2 rds); ac losses 0/1.	Fox, 1979, p.183
174	2/24/69	Phan Rang	Destroy aircraft	Standoff attack (10 rds); ac losses 0/0.	Fox, 1979, p.183
175	2/24/69	Nha Trang	Harass	Standoff attack (5 rds); ac losses 0/0.	Fox, 1979, p.183
176	2/25/69	Da Nang	Harass	Standoff attack (3 rds); ac losses 0/0.	Fox, 1979, p.183
177	2/25/69	Pleiku	Harass	Standoff attack (1 rd); ac losses 0/0.	Fox, 1979, p.184
178	3/15/69	Phan Rang	Destroy aircraft	Standoff attack (34 rds); ac losses 0/0.	Fox, 1979, p.184
179	3/15/69	Phan Rang	Destroy aircraft	Standoff attack (7 rds); ac losses 0/0.	Fox, 1979, p.184
180	3/16/69	Phan Rang	Harass	Standoff attack (5 rds); ac losses 0/0.	Fox, 1979, p.184
181	3/19/69	Phan Rang	Destroy aircraft	Standoff attack (36 rds); ac losses 0/0.	Fox, 1979, p.184
182	3/21/69	Cam Ranh Bay	Destroy aircraft	Standoff attack (7 rds); ac losses 0/0.	Fox, 1979, p.184
183	3/21/69	Da Nang	Harass	Standoff attack (5 rds); ac losses 0/0.	Fox, 1979, p.184
184	3/21/69	Pleiku	Harass	Standoff attack (3 rds); ac losses 0/0.	Fox, 1979, p.184
185	3/21/69	Phan Rang	Destroy aircraft	Standoff attack (25 rds); ac losses 0/0.	Fox, 1979, p.184
186	3/24/69	Phan Rang	Destroy aircraft	Standoff attack (41 rds); ac losses 0/0.	Fox, 1979, p.184

Table B.3—continued

Event	Date	Location	Objective	Description (aircraft [ac] losses destroyed/damaged)	Sources
187	3/24/69	Da Nang	Destroy aircraft	Standoff attack (14 rds); ac losses 0/0.	Fox, 1979, p.184
188	3/27/69	Pleiku	Harass	Standoff attack (1 rd); ac losses 0/0.	Fox, 1979, p.184
189	3/29/69	Bien Hoa	Harass	Standoff attack (2 rds); ac losses 0/0.	Fox, 1979, p.184
190	3/31/69	Bien Hoa	Harass	Standoff attack (2 rds); ac losses 0/0.	Fox, 1979, p.184
191	4/13/69	Phan Rang	Destroy aircraft	Standoff attack (13 rds); ac losses 0/0.	Fox, 1979, p.184
192	4/16/69	Phu Cat	Destroy aircraft	Sapper attack; ac losses 0/0.	Fox, 1979, p.184
193	4/17/69	Da Nang	Harass	Standoff attack (2 rds); ac losses 0/0.	Fox, 1979, p.185
194	4/20/69	Da Nang	Destroy aircraft	Standoff attack (3 rds); ac losses 0/1.	Fox, 1979, p.185
195	4/21/69	Nha Trang	Destroy aircraft	Standoff attack (6 rds); ac losses 0/4.	Fox, 1979, p.185
196	4/21/69	Phan Rang	Harass	Standoff attack (5 rds); ac losses 0/0.	Fox, 1979, p.185
197	4/24/69	Da Nang	Harass	Standoff attack (2 rds); ac losses 0/0.	Fox, 1979, p.185
198	4/25/69	Pleiku	Destroy aircraft	Standoff attack (1 rd); ac losses 0/1.	Fox, 1979, p.185
199	5/11/69	Binh Thuy	Destroy aircraft	Standoff attack (11 rds); ac losses 0/0.	Fox, 1979, p.185
200	5/11/69	Pleiku	Harass	Standoff attack (3 rds); ac losses 0/0.	Fox, 1979, p.185
201	5/12/69	Phan Rang	Destroy aircraft	Standoff attack (30 rds); ac losses 0/1.	Fox, 1979, p.185
202	5/12/69	Bien Hoa	Destroy aircraft	Standoff attack (5 rds); ac losses 0/3.	Fox, 1979, p.185
203	5/12/69	Da Nang	Destroy aircraft	Standoff attack (3 rds); ac losses 0/1.	Fox, 1979, p.185
204	5/12/69	Phan Rang	Destroy aircraft	Standoff attack (9 rds); ac losses 0/1.	Fox, 1979, p.185
205	5/12/69	Tan Son Nhut	Harass	Standoff attack (3 rds); ac losses 0/0.	Fox, 1979, p.185
206	5/14/69	Da Nang	Harass	Standoff attack (1 rd); ac losses 0/0.	Fox, 1979, p.185
207	5/16/69	Phan Rang	Destroy aircraft	Standoff attack (22 rds); ac losses 0/0.	Fox, 1979, p.185

Table B.3—continued

Event	Date	Location	Objective	Description (aircraft [ac] losses destroyed/damaged)	Sources
208	5/17/69	Da Nang	Destroy aircraft	Standoff attack (2 rds); ac losses 0/4.	Fox, 1979, p.185
209	5/21/69	Bien Hoa	Harass	Standoff attack (2 rds); ac losses 0/0.	Fox, 1979, p.186
210	5/22/69	Phan Rang	Destroy aircraft	Standoff attack (18 rds); ac losses 0/0.	Fox, 1979, p.186
211	5/22/69	Phan Rang	Harass	Standoff attack (1 rd); ac losses 0/0.	Fox, 1979, p.186
212	5/23/69	Bien Hoa	Harass	Standoff attack (3 rds); ac losses 0/0.	Fox, 1979, p.186
213	5/28/69	Bien Hoa	Harass	Standoff attack (4 rds); ac losses 0/0.	Fox, 1979, p.186
214	5/31/69	Nha Trang	Destroy aircraft	Standoff attack (10 rds); ac losses 0/0.	Fox, 1979, p.186
215	6/5/69	Bien Hoa	Destroy aircraft	Standoff attack (4 rds); ac losses 0/1.	Fox, 1979, p.186
216	6/6/69	Binh Thuy	Destroy aircraft	Standoff attack (11 rds); ac losses 0/0.	Fox, 1979, p.186
217	6/6/69	Phan Rang	Destroy aircraft	Standoff attack (15 rds); ac losses 0/1.	Fox, 1979, p.186
218	6/6/69	Bien Hoa	Destroy aircraft	Standoff attack (36 rds); ac losses 0/2.	Fox, 1979, p.186
219	6/7/69	Da Nang	Destroy aircraft	Standoff attack (20 rds); ac loses 2/12.	Fox, 1979, p.186
220	6/7/69	Pleiku	Harass	Standoff attack (1 rd); ac losses 0/0.	Fox, 1979, p.186
221	6/7/69	Phan Rang	Harass	Standoff attack (3 rds); ac losses 0/0.	Fox, 1979, p.186
222	6/9/69	Bien Hoa	Harass	Standoff attack (3 rds); ac losses 0/0.	Fox, 1979, p.186
223	6/11/69	Phan Rang	Destroy aircraft	Standoff attack (17 rds); ac losses 0/0.	Fox, 1979, p.186
224	6/12/69	Pleiku	Harass	Standoff attack (1 rd); ac losses 0/0.	Fox, 1979, p.186
225	6/12/69	Bien Hoa	Destroy aircraft	Standoff attack (30 rds); ac losses 0/0.	Fox, 1979, p.187
226	6/16/69	Bien Hoa	Destroy aircraft	Standoff attack (4 rds); ac losses 0/1.	Fox, 1979, p.187
227	6/17/69	Phu Cat	Destroy aircraft	Standoff attack (18 rds); ac losses 0/0.	Fox, 1979, p.187

Table B.3—continued

Event	Date	Location	Objective	Description (aircraft [ac] losses destroyed/damaged)	Sources
228	6/18/69	Bien Hoa	Destroy aircraft	Standoff attack (9 rds); ac losses 0/0.	Fox, 1979, p.187
229	6/18/69	Phan Rang	Destroy aircraft	Standoff attack (14 rds); ac losses 0/0.	Fox, 1979, p.187
230	6/20/69	Phan Rang	Destroy aircraft	Standoff attack (4 rds); ac losses 0/1.	Fox, 1979, p.187
231	6/20/69	Bien Hoa	Destroy aircraft	Standoff attack (8 rds); ac losses 0/0.	Fox, 1979, p.187
232	6/29/69	Tan Son Nhut	Harass	Standoff attack (3 rds); ac losses 0/0.	Fox, 1979, p.187
233	7/8/69	Cam Ranh Bay	Destroy aircraft	Standoff attack (12 rds); ac losses 0/0.	Fox, 1979, p.187
234	7/10/69	Bien Hoa	Harass	Standoff attack (4 rds); ac losses 0/0.	Fox, 1979, p.187
235	7/10/69	Binh Thuy	Harass	Standoff attack (1 rd); ac losses 0/0.	Fox, 1979, p.187
236	7/15/69	Phan Rang	Harass	Standoff attack (3 rds); ac losses 0/0.	Fox, 1979, p.187
237	7/19/69	Phan Rang	Destroy aircraft	Standoff attack (11 rds); ac losses 0/0.	Fox, 1979, p.187
238	7/20/69	Bien Hoa	Destroy aircraft	Standoff attack (29 rds); ac losses 0/0.	Fox, 1979, p.187
239	7/20/69	Phan Rang	Harass	Standoff attack (3 rds); ac losses 0/0.	Fox, 1979, p.187
240	7/28/69	Ubon, Thailand	Destroy aircraft	Sapper attack; ac losses 0/2.	USAF, 1973
241	8/7/69	Cam Ranh Bay	Destroy aircraft	Standoff attack (22 rds); ac losses 0/10.	Fox, 1979, p.187
242	8/12/69	Bien Hoa	Destroy aircraft	Standoff attack (8 rds); ac losses 0/0.	Fox, 1979, p.188
243	8/13/69	Da Nang	Harass	Standoff attack (5 rds); ac losses 0/0.	Fox, 1979, p.188
244	8/22/69	Da Nang	Destroy aircraft	Standoff attack (10 rds); ac losses 0/0.	Fox, 1979, p.188
245	9/2/69	Pleiku	Destroy aircraft	Standoff attack (1 rd); ac losses 0/3.	Fox, 1979, p.188
246	9/4/69	Phan Rang	Destroy aircraft	Standoff attack (18 rds); ac losses 0/3.	Fox, 1979, p.188
247	9/5/69	Bien Hoa	Harass	Standoff attack (4 rds); ac losses 0/0.	Fox, 1979, p.188
248	9/6/69	Cam Ranh Bay	Destroy aircraft	Standoff attack (4 rds); ac losses 0/9.	Fox, 1979, p.188

Table B.3—continued

Event	Date	Location	Objective	Description (aircraft [ac] losses destroyed/damaged)	Sources
249	9/6/69	Da Nang	Destroy aircraft	Standoff attack (8 rds); ac losses 0/0.	Fox, 1979, p.188
250	9/6/69	Bien Hoa	Destroy aircraft	Standoff attack (18 rds); ac losses 0/1.	Fox, 1979, p.188
251	9/13/69	Phan Rang	Harass	Standoff attack (5 rds); ac losses 0/0.	Fox, 1979, p.188
252	9/20/69	Phan Rang	Harass	Standoff attack (3 rds); ac losses 0/0.	Fox, 1979, p.188
253	10/11/69	Nha Trang	Destroy aircraft	Standoff attack (10 rds); ac losses 0/2.	Fox, 1979, p.188
254	10/12/69	Nha Trang	Harass	Standoff attack (3 rds); ac losses 0/0.	Fox, 1979, p.188
255	10/25/69	Pleiku	Harass	Standoff attack (3 rds); ac losses 0/0.	Fox, 1979, p.188
256	11/4/69	Phan Rang	Harass	Standoff attack (3 rds); ac losses 0/0.	Fox, 1979, p.188
257	11/4/69	Phan Rang	Harass	Standoff attack (2 rds); ac losses 0/0.	Fox, 1979, p.188
258	11/9/69	Phan Rang	Harass	Standoff attack (2 rds); ac losses 0/0.	Fox, 1979, p.189
259	11/14/69	Cam Ranh Bay	Destroy aircraft	Standoff attack (8 rds); ac losses 0/1.	Fox, 1979, p.189
260	11/16/69	Pleiku	Harass	Standoff attack (1 rd); ac losses 0/0.	Fox, 1979, p.189
261	11/16/69	Phan Rang	Harass	Standoff attack (1 rd); ac losses 0/0.	Fox, 1979, p.189
262	11/21/69	Phan Rang	Harass	Standoff attack (1 rd); ac losses 0/0.	Fox, 1979, p.189
263	11/25/69	Bien Hoa	Harass	Standoff attack (3 rds); ac losses 0/0.	Fox, 1979, p.189
264	12/3/69	Phan Rang	Harass	Standoff attack (1 rd); ac losses 0/0.	Fox, 1979, p.189
265	12/7/69	Cam Ranh Bay	Harass	Standoff attack (3 rds); ac losses 0/0.	Fox, 1979, p.189
266	12/11/69	Da Nang	Destroy aircraft	Standoff attack (4 rds); ac losses 0/1.	Fox, 1979, p.189
267	12/12/69	Bien Hoa	Destroy aircraft	Standoff attack (11 rds); ac losses 1/0.	Fox, 1979, p.189
268	12/14/69	Phan Rang	Harass	Standoff attack (3 rds); ac losses 0/0.	Fox, 1979, p.189
269	12/19/69	Tan Son Nhut	Harass	Standoff attack (4 rds); ac losses 0/0.	Fox, 1979, p.189

Table B.3—continued

Event	Date	Location	Objective	Description (aircraft [ac] losses destroyed/damaged)	Sources
270	1/4/70	Phu Cat	Destroy aircraft	Standoff attack (3 rds); ac losses 0/5.	Fox, 1979, p.189
271	1/4/70	Phu Cat	Harass	Standoff attack (2 rds); ac losses 0/0.	Fox, 1979, p.189
272	1/5/70	Phan Rang	Harass	Standoff attack (3 rds); ac losses 0/0.	Fox, 1979, p.189
273	1/6/70	Cam Ranh Bay	Harass	Standoff attack (2 rds); ac losses 0/0.	Fox, 1979, p.190
274	1/7/70	Cam Ranh Bay	Harass	Standoff attack (2 rds); ac losses 0/0.	Fox, 1979, p.190
275	1/9/70	Cam Ranh Bay	Harass	Standoff attack (1 rd); ac losses 0/0.	Fox, 1979, p.190
276	1/12/70	Ubon,Thailand	Destroy aircraft	Sappers attempted to destroy AC-130 gunships; detected and killed. No aircraft damaged.	USAF, 1973
277	1/13/70	Cam Ranh Bay	Harass	Standoff attack (2 rds); ac losses 0/0.	Fox, 1979, p.190
278	1/20/70	Phan Rang	Harass	Standoff attack (1 rd); ac losses 0/0.	Fox, 1979, p.190
279	1/21/70	Bien Hoa	Destroy aircraft	Standoff attack (8 rds); ac losses 0/3.	Fox, 1979, p.190
280	1/25/70	Phan Rang	Harass	Standoff attack (2 rds); ac losses 0/0.	Fox, 1979, p.190
281	2/2/70	Cam Ranh Bay	Harass	Standoff attack (1 rd); ac losses 0/0.	Fox, 1979, p.190
282	2/2/70	Phu Cat	Destroy aircraft	Standoff attack (10 rds); ac losses 0/0.	Fox, 1979, p.190
283	2/4/70	Bien Hoa	Harass	Standoff attack (4 rds); ac losses 0/0.	Fox, 1979, p.190
284	2/11/70	Phan Rang	Destroy aircraft	Sapper attack; ac losses 0/0.	Fox, 1979, p.190
285	2/16/70	Phan Rang	Destroy aircraft	Standoff attack (8 rds); ac losses 0/0.	Fox, 1979, p.190
286	2/21/70	Phan Rang	Harass	Standoff attack (5 rds); ac losses 0/0.	Fox, 1979, p.190
287	2/27/70	Bien Hoa	Destroy aircraft	Standoff attack (6 rds); ac losses 0/6.	Fox, 1979, p.190
288	3/4/70	Phan Rang	Harass	Standoff attack (1 rd); ac losses 0/0.	Fox, 1979, p.190
289	3/7/70	Cam Ranh Bay	Destroy aircraft	Standoff attack (8 rds); ac losses 0/0.	Fox, 1979, p.190

Table B.3—continued

Event	Date	Location	Objective	Description (aircraft [ac] losses destroyed/damaged)	Sources
290	3/7/70	Cam Ranh Bay	Harass	Standoff attack (3 rds); ac losses 0/0.	Fox, 1979, p.191
291	3/12/70	Cam Ranh Bay	Harass	Standoff attack (4 rds); ac losses 0/0.	Fox, 1979, p.191
292	3/14/70	Phan Rang	Destroy aircraft	Standoff attack (7 rds); ac losses 0/0.	Fox, 1979, p.191
293	4/1/70	Phan Rang	Destroy aircraft	Standoff attack (12 rds); ac losses 0/0.	Fox, 1979, p.191
294	4/1/70	Bien Hoa	Destroy aircraft	Standoff attack (5 rds); ac losses 0/1.	Fox, 1979, p.191
295	4/1/70	Phan Rang	Harass	Standoff attack (2 rds); ac losses 0/0.	Fox, 1979, p.191
296	4/4/70	Bien Hoa	Harass	Standoff attack (2 rds); ac losses 0/0.	Fox, 1979, p.191
297	4/4/70	Phu Cat	Destroy aircraft	Sapper attack; ac losses 0/0.	Fox, 1979, p.191
298	4/5/70	Phan Rang	Harass	Standoff attack (1 rd); ac losses 0/0.	Fox, 1979, p.191
299	4/6/70	Nha Trang	Destroy aircraft	Standoff attack (11 rds); ac losses 0/0.	Fox, 1979, p.191
300	4/7/70	Phan Rang	Destroy aircraft	Standoff attack (6 rds); ac losses 0/0.	Fox, 1979, p.191
301	4/8/70	Da Nang	Harass	Standoff attack (4 rds); ac losses 0/0.	Fox, 1979, p.191
302	4/8/70	Cam Ranh Bay	Harass	Standoff attack (4 rds); ac losses 0/0.	Fox, 1979, p.191
303	4/9/70	Phan Rang	Harass	Standoff attack (1 rd); ac losses 0/0.	Fox, 1979, p.191
304	4/19/70	Cam Ranh Bay	Harass	Standoff attack (3 rds); ac losses 0/0.	Fox, 1979, p.191
305	4/20/70	Phan Rang	Harass	Standoff attack (1 rd); ac losses 0/0.	Fox, 1979, p.191
306	5/3/70	Phan Rang	Destroy aircraft	Standoff & sapper attack (12 rds); ac losses 0/0.	Fox, 1979, p.192
307	5/3/70	Bien Hoa	Destroy aircraft	Standoff attack (6 rds); ac losses 0/0.	Fox, 1979, p.192
308	5/3/70	Bien Hoa	Destroy aircraft	Standoff attack (4 rds); ac losses 1/0.	Fox, 1979, p.192
309	5/3/70	Bien Hoa	Destroy aircraft	Standoff attack (7 rds); ac losses 0/0.	Fox, 1979, p.192

Table B.3—continued

Event	Date	Location	Objective	Description (aircraft [ac] losses destroyed/damaged)	Sources
310	5/4/70	Bien Hoa	Harass	Standoff attack (3 rds); ac losses 0/0.	Fox, 1979, p.192
311	5/6/70	Phan Rang	Destroy aircraft	Standoff attack (6 rds); ac losses 0/0.	Fox, 1979, p.192
312	5/7/70	Pleiku	Harass	Standoff attack (5 rds); ac losses 0/0.	Fox, 1979, p.192
313	5/7/70	Phan Rang	Harass	Standoff attack (1 rd); ac losses 0/0.	Fox, 1979, p.192
314	5/8/70	Tuy Hoa	Destroy aircraft	Standoff attack (32 rds); ac losses 0/0.	Fox, 1979, p.192
315	5/8/70	Cam Ranh Bay	Destroy aircraft	Standoff attack (26 rds); ac losses 0/0.	Fox, 1979, p.192
316	5/8/70	Phu Cat	Harass	Standoff attack (4 rds); ac losses 0/0.	Fox, 1979, p.192
317	5/12/70	Cam Ranh Bay	Harass	Standoff attack (3 rds); ac losses 0/0.	Fox, 1979, p.192
318	5/15/70	Pleiku	Harass	Standoff attack (4 rds); ac losses 0/0.	Fox, 1979, p.192
319	5/16/70	Phan Rang	Destroy aircraft	Standoff attack (12 rds); ac losses 0/0.	Fox, 1979, p.192
320	5/19/70	Cam Ranh Bay	Harass	Standoff attack (5 rds); ac losses 0/0.	Fox, 1979, p.192
321	5/19/70	Pleiku	Destroy aircraft	Standoff attack (4 rds); ac losses 1/2.	Fox, 1979, p.192
322	5/21/70	Phu Cat	Destroy aircraft	Standoff attack (6 rds); ac losses 0/3.	Fox, 1979, p.193
323	5/21/70	Da Nang	Destroy aircraft	Standoff attack (3 rds); ac losses 0/3.	Fox, 1979, p.193
324	5/26/70	Pleiku	Destroy aircraft	Standoff attack (4 rds); ac losses 0/2.	Fox, 1979, p.193
325	5/30/70	Phan Rang	Harass	Standoff attack (1 rd); ac losses 0/0.	Fox, 1979, p.193
326	6/4/70	Phu Cat	Harass	Standoff attack (5 rds); ac losses 0/0.	Fox, 1979, p.193
327	6/4/70	Cam Ranh Bay	Harass	Standoff attack (4 rds); ac losses 0/0.	Fox, 1979, p.193
328	6/4/70	Nha Trang	Harass	Standoff attack (2 rds); ac losses 0/0.	Fox, 1979, p.193
329	6/4/70	Nha Trang	Harass	Standoff attack (3 rds); ac losses 0/0.	Fox, 1979, p.193
330	6/6/70	Phan Rang	Harass	Standoff attack (2 rds); ac losses 0/0.	Fox, 1979, p.193

Table B.3—continued

Event	Date	Location	Objective	Description (aircraft [ac] losses destroyed/damaged)	Sources
331	6/6/70	Cam Ranh Bay	Harass	Standoff attack (4 rds); ac losses 0/0.	Fox, 1979, p.193
332	6/7/70	Nha Trang	Harass	Standoff attack (2 rds); ac losses 0/0.	Fox, 1979, p.193
333	6/7/70	Nha Trang	Harass	Standoff attack (3 rds); ac losses 0/0.	Fox, 1979, p.193
334	6/7/70	Binh Thuy	Destroy aircraft	Standoff attack (6 rds); ac losses 0/0.	Fox, 1979, p.193
335	6/10/70	Phan Rang	Harass	Standoff attack (3 rds); ac losses 0/0.	Fox, 1979, p.193
336	6/10/70	Cam Ranh Bay	Harass	Standoff attack (2 rds); ac losses 0/0.	Fox, 1979, p.193
337	6/11/70	Bien Hoa	Harass	Standoff attack (2 rds); ac losses 0/0.	Fox, 1979, p.193
338	6/12/70	Cam Ranh Bay	Destroy aircraft	Sapper attack; ac losses 0/1.	Fox, 1979, p.194
339	6/21/70	Da Nang	Harass	Standoff attack (3 rds); ac losses 0/0.	Fox, 1979, p.194
340	6/25/70	Bien Hoa	Harass	Standoff attack (1 rd); ac losses 0/0.	Fox, 1979, p.194
341	7/2/70	Phan Rang	Harass	Standoff attack (2 rds); ac losses 0/0.	Fox, 1979, p.194
342	7/4/70	Tuy Hoa	Destroy aircraft	Standoff attack (20 rds); ac losses 0/0.	Fox, 1979, p.194
343	7/7/70	Binh Thuy	Destroy aircraft	Standoff attack (2 rds); ac losses 0/1.	Fox, 1979, p.194
344	7/9/70	Cam Ranh Bay	Destroy aircraft	Standoff & sapper (6 rds); ac losses 0/0.	Fox, 1979, p.194
345	7/9/70	Phan Rang	Harass	Standoff attack (2 rds); ac losses 0/0.	Fox, 1979, p.194
346	7/21/70	Binh Thuy	Harass	Standoff attack (1 rd); ac losses 0/0.	Fox, 1979, p.194
347	7/21/70	Phan Rang	Harass	Standoff attack (1 rd); ac losses 0/0.	Fox, 1979, p.194
348	8/1/70	Binh Thuy	Harass	Standoff attack (4 rds); ac losses 0/0.	Fox, 1979, p.194
349	8/5/70	Phan Rang	Harass	Standoff attack (1 rd); ac losses 0/0.	Fox, 1979, p.194
350	8/7/70	Cam Ranh Bay	Harass	Standoff attack (3 rds); ac losses 0/0.	Fox, 1979, p.194

Table B.3—continued

Event	Date	Location	Objective	Description (aircraft [ac] losses destroyed/damaged)	Sources
351	8/12/70	Cam Ranh Bay	Harass	Standoff attack (3 rds); ac losses 0/0.	Fox, 1979, p.194
352	8/22/70	Phan Rang	Harass	Standoff attack (1 rd); ac losses 0/0.	Fox, 1979, p.194
353	8/30/70	Cam Ranh Bay	Destroy aircraft	Standoff & sapper attack (7 rds) destroyed 460,000 gallons of fuel and 2.25 million gallons of storage capacity.	Fox, 1979, p.195
354	8/30/70	Phu Cat	Destroy aircraft	Standoff attack (6 rds); ac losses 0/0.	Fox, 1979, p.195
355	8/30/70	Nha Trang	Harass	Standoff attack (3 rds); ac losses 0/0.	Fox, 1979, p.195
356	8/31/70	Phan Rang	Harass	Standoff attack (1 rd); ac losses 0/0.	Fox, 1979, p.195
357	9/1/70	Da Nang	Destroy aircraft	Standoff attack (8 rds); ac losses 0/1.	Fox, 1979, p.195
358	9/4/70	Pleiku	Harass	Standoff attack (2 rds); ac losses 0/0.	Fox, 1979, p.195
359	9/16/70	Binh Thuy	Harass	Standoff attack (3 rds); ac losses 0/0.	Fox, 1979, p.195
360	10/4/70	Phan Rang	Harass	Standoff attack (2 rds); ac losses 0/0.	Fox, 1979, p.195
361	10/5/70	Phu Cat	Harass	Standoff attack (2 rds); ac losses 0/0.	Fox, 1979, p.195
362	10/12/70	Da Nang	Harass	Standoff attack (2 rds); ac losses 0/0.	Fox, 1979, p.195
363	10/21/70	Da Nang	Harass	Standoff attack (1 rd); ac losses 0/0.	Fox, 1979, p.195
364	11/8/70	Phan Rang	Harass	Standoff attack (1 rd); ac losses 0/0.	Fox, 1979, p.195
365	11/17/70	Bien Hoa	Destroy aircraft	Standoff attack (28 rds); ac losses 0/0.	Fox, 1979, p.195
366	11/21/70	Pleiku	Harass	Standoff attack (5 rds); ac losses 0/0.	Fox, 1979, p.196
367	11/23/70	Pleiku	Harass	Standoff attack (3 rds); ac losses 0/0.	Fox, 1979, p.196
368	11/24/70	Pleiku	Destroy aircraft	Standoff attack (17 rds); ac losses 0/0.	Fox, 1979, p.196
369	11/25/70	Pleiku	Harass	Standoff attack (1 rd); ac losses 0/0.	Fox, 1979, p.196

Table B.3—continued

Event	Date	Location	Objective	Description (aircraft [ac] losses destroyed/damaged)	Sources
370	11/29/70	Phan Rang	Harass	Standoff attack (2 rds); ac losses 0/0.	Fox, 1979, p.196
371	12/1/70	Cam Ranh Bay	Harass	Standoff attack (3 rds); ac losses 0/0.	Fox, 1979, p.196
372	12/2/70	Phu Cat	Harass	Standoff attack (3 rds); ac losses 0/0.	Fox, 1979, p.196
373	12/6/70	Cam Ranh Bay	Harass	Standoff attack (4 rds); ac losses 0/0.	Fox, 1979, p.196
374	12/16/70	Bien Hoa	Harass	Standoff attack (1 rd); ac losses 0/0.	Fox, 1979, p.196
375	12/21/70	Da Nang	Harass	Standoff attack (1 rd); ac losses 0/0.	Fox, 1979, p.196
376	12/29/70	Pleiku	Harass	Standoff attack (2 rds); ac losses 0/0.	Fox, 1979, p.196
377	1/22/71	Bien Hoa	Harass	Standoff attack (1 rd); ac losses 0/0.	Fox, 1979, p.196
378	2/1/71	Da Nang	Destroy aircraft	Standoff attack (8 rds); ac losses 0/2.	Fox, 1979, p.196
379	2/1/71	Nha Trang	Destroy aircraft	Standoff attack (no. rds not recorded); ac losses 0/0.	Fox, 1979, p.196
380	2/1/71	Phu Cat	Destroy aircraft	Standoff attack (6 rds); ac losses 0/0.	Fox, 1979, p.196
381	2/1/71	Nha Trang	Destroy aircraft	Standoff attack (no. rds not recorded); ac losses 0/0.	Fox, 1979, p.197
382	2/21/71	Da Nang	Destroy aircraft	Standoff attack (6 rds); ac losses 1/3.	Fox, 1979, p.197
383	2/21/71	Phu Cat	Destroy aircraft	Standoff attack (4 rds); ac losses 0/3.	Fox, 1979, p.197
384	2/22/71	Phu Cat	Harass	Standoff attack (4 rds); ac losses 0/0.	Fox, 1979, p.197
385	2/24/71	Phu Cat	Harass	Standoff attack (2 rds); ac losses 0/0.	Fox, 1979, p.197
386	2/24/71	Cam Ranh Bay	Destroy aircraft	Sapper attack; ac losses 0/0.	Fox, 1979, p.197
387	2/25/71	Nha Trang	Harass	Standoff attack (2 rds); ac losses 0/0.	Fox, 1979, p.197
388	2/25/71	Nha Trang	Destroy aircraft	Standoff attack (2 rds); ac losses 0/2.	Fox, 1979, p.197
389	2/28/71	Pleiku	Destroy aircraft	Sapper attack; ac losses 0/0.	Fox, 1979, p.197

Table B.3—continued

Event	Date	Location	Objective	Description (aircraft [ac] losses destroyed/damaged)	Sources
390	2/28/71	Cam Ranh Bay	Destroy aircraft	Standoff attack (6 rds); ac losses 0/0.	Fox, 1979, p.197
391	3/4/71	Da Nang	Destroy aircraft	Standoff attack (10 rds); ac losses 0/0.	Fox, 1979, p.197
392	3/16/71	Bien Hoa	Harass	Standoff attack (5 rds); ac losses 0/0.	Fox, 1979, p.197
393	3/19/71	Cam Ranh Bay	Harass	Standoff attack (3 rds); ac losses 0/0.	Fox, 1979, p.197
394	3/20/71	Cam Ranh Bay	Destroy aircraft	Standoff attack (5 rds); ac losses 0/1.	Fox, 1979, p.197
395	3/29/71	Da Nang	Destroy aircraft	Standoff attack (2 rds); ac losses 0/2.	Fox, 1979, p.197
396	3/29/71	Da Nang	Harass	Standoff attack (2 rds); ac losses 0/0.	Fox, 1979, p.197
397	3/31/71	Pleiku	Destroy aircraft	Standoff & sapper attack (12 rds); ac losses 0/15.	Fox, 1979, p.198
398	4/4/71	Pleiku	Harass	Standoff attack (3 rds); ac losses 0/0.	Fox, 1979, p.198
399	4/5/71	Da Nang	Destroy aircraft	Standoff attack (3 rds); ac losses 0/1.	Fox, 1979, p.198
400	4/9/71	Da Nang	Harass	Standoff attack (1 rd); ac losses 0/0.	Fox, 1979, p.198
401	4/16/71	Cam Ranh Bay	Harass	Standoff attack (3 rds); ac losses 0/0.	Fox, 1979, p.198
402	4/25/71	Cam Ranh Bay	Harass	Standoff attack (3 rds); ac losses 0/0.	Fox, 1979, p.198
403	4/26/71	Da Nang	Harass	Standoff attack (1 rd); ac losses 0/0.	Fox, 1979, p.198
404	4/27/71	Da Nang	Harass	Standoff attack (3 rds); ac losses 0/0.	Fox, 1979, p.198
405	4/27/71	Binh Thuy	Harass	Standoff attack (1 rd); ac losses 0/0.	Fox, 1979, p.198
406	5/1/71	Binh Thuy	Harass	Standoff attack (3 rds); ac losses 0/0.	Fox, 1979, p.198
407	5/5/71	Da Nang	Destroy aircraft	Standoff attack (2 rds); ac losses 0/1.	Fox, 1979, p.198
408	5/6/71	Pleiku	Harass	Standoff attack (3 rds); ac losses 0/0.	Fox, 1979, p.198
409	5/13/71	Binh Thuy	Harass	Standoff attack (3 rds); ac losses 0/0.	Fox, 1979, p.198
410	5/23/71	Cam Ranh Bay	Destroy aircraft	Sapper attack; ac losses 0/0.	Fox, 1979, p.198

Table B.3—continued

Event	Date	Location	Objective	Description (aircraft [ac] losses destroyed/damaged)	Sources
411	5/30/71	Da Nang	Destroy aircraft	Standoff attack (7 rds); ac losses 0/0.	Fox, 1979, p.198
412	6/5/71	Da Nang	Destroy aircraft	Standoff attack (6 rds); ac losses 0/0.	Fox, 1979, p.198
413	6/7/71	Da Nang	Harass	Standoff attack (2 rds); ac losses 0/0.	Fox, 1979, p.199
414	6/11/71	Cam Ranh Bay	Harass	Standoff attack (3 rds); ac losses 0/0.	Fox, 1979, p.199
415	7/5/71	Da Nang	Harass	Standoff attack (3 rds); ac losses 0/0.	Fox, 1979, p.199
416	7/27/71	Phan Rang	Destroy aircraft	Standoff attack (7 rds); ac losses 0/0.	Fox, 1979, p.199
417	8/16/71	Bien Hoa	Harass	Standoff attack (2 rds); ac losses 0/0.	Fox, 1979, p.199
418	8/25/71	Da Nang	Harass	Standoff attack (2 rds); ac losses 0/0.	Fox, 1979, p.199
419	8/25/71	Cam Ranh Bay	Destroy aircraft	Standoff & sapper attack (5 rds); destroyed 6,000 tons of munitions.	USAF, 1971b
420	8/28/71	Tan Son Nhut	Harass	Standoff attack (3 rds); ac losses 0/0.	Fox, 1979, p.199
421	8/29/71	Pleiku	Destroy aircraft	Standoff attack (6 rds); ac losses 0/0.	Fox, 1979, p.199
422	9/13/71	Cam Ranh Bay	Destroy aircraft	Sapper attack; ac losses 0/0.	Fox, 1979, p.199
423	9/21/71	Pleiku	Harass	Standoff attack (2 rds); ac losses 0/0.	Fox, 1979, p.199
424	9/25/71	Bien Hoa	Harass	Standoff attack (3 rds); ac losses 0/0.	Fox, 1979, p.199
425	9/25/71	Phan Rang	Harass	Standoff attack (3 rds); ac losses 0/0.	Fox, 1979, p.199
426	9/29/71	Bien Hoa	Harass	Standoff attack (1 rd); ac losses 0/0.	Fox, 1979, p.199
427	10/2/71	Da Nang	Harass	Standoff attack (4 rds); ac losses 0/0.	Fox, 1979, p.199
428	10/3/71	Bien Hoa	Harass	Standoff attack (3 rds); ac losses 0/0.	Fox, 1979, p.200
429	11/9/71	Phan Rang	Harass	Standoff attack (2 rds); ac losses 0/0.	Fox, 1979, p.200
430	11/15/71	Cam Ranh Bay	Harass	Standoff attack (4 rds); ac losses 0/0.	Fox, 1979, p.200

Table B.3—continued

Event	Date	Location	Objective	Description (aircraft [ac] losses destroyed/damaged)	Sources
431	11/25/71	Bien Hoa	Harass	Standoff attack (3 rds); ac losses 0/0.	Fox, 1979, p.200
432	1972	Unknown location, Thailand	Destroy aircraft	USAF AC-119 destroyed by enemy ground forces.	USAF , 1974, p.27; Francillon, 1987, Table 2D
433	1/3/72	Da Nang	Destroy aircraft	Standoff attack (6 rds); ac losses 0/2.	Fox, 1979, p.200
434	1/10/72	U-Tapao, Thailand	Destroy aircraft	Sapper attack; ac losses 0/3.	USAF, 1973
435	1/12/72	Bien Hoa	Destroy aircraft	Sapper attack destroyed $400,000 worth of munitions.	Fox, 1979, p.200
436	1/16/72	Cam Ranh Bay	Harass	Standoff attack (4 rds); ac losses 0/0.	Fox, 1979, p.200
437	2/5/72	Phan Rang	Harass	Standoff attack (1 rd); ac losses 0/0.	Fox, 1979, p.200
438	2/9/72	Da Nang	Destroy aircraft	Standoff attack (28 rds); ac losses 0/2.	Fox, 1979, p.200
439	2/21/72	Bien Hoa	Harass	Standoff attack (3 rds); ac losses 0/0.	Fox, 1979, p.200
440	2/21/72	Phan Rang	Harass	Standoff attack (2 rds); ac losses 0/0.	Fox, 1979, p.200
441	3/6/72	Cam Ranh Bay	Destroy aircraft	Standoff attack (3 rds); ac losses 0/1.	Fox, 1979, p.200
442	4/7/72	Bien Hoa	Harass	Standoff attack (4 rds); ac losses 0/0.	Fox, 1979, p.201
443	4/13/72	Cam Ranh Bay	Destroy aircraft	Standoff attack (7 rds); ac losses 0/0.	Fox, 1979, p.201
444	4/13/72	Da Nang	Destroy aircraft	Standoff attack (24 rds); ac losses 1/9.	Fox, 1979, p.201
445	4/14/72	Tan Son Nhut	Harass	Standoff attack (4 rds); ac losses 0/0.	Fox, 1979, p.201
446	4/16/72	Da Nang	Destroy aircraft	Standoff attack (20 rds); ac losses 0/1.	Fox, 1979, p.201
447	4/24/72	Da Nang	Destroy aircraft	Standoff attack (13 rds); ac losses 0/0.	Fox, 1979, p.201
448	4/25/72	Da Nang	Destroy aircraft	Standoff attack (6 rds); ac losses 0/0.	Fox, 1979, p.201

Table B.3—continued

Event	Date	Location	Objective	Description (aircraft [ac] losses destroyed/damaged)	Sources
449	5/7/72	Da Nang	Destroy aircraft	Standoff attack (16 rds); ac losses 1/2.	Fox, 1979, p.201
450	5/12/72	Da Nang	Destroy aircraft	Standoff attack (18 rds); ac losses 0/3.	Fox, 1979, p.201
451	5/14/72	Da Nang	Destroy aircraft	Standoff attack (18 rds); ac losses 0/2.	Fox, 1979, p.201
452	5/17/72	Kontum	Deny defender use of airfield	Standoff attack; destroyed 1 C-130 and damaged 2 C-130s & 2 C-123s.	Berger, 1984, p.177
453	5/20/72	Kontum	Deny defender use of airfield	Standoff attack; destroyed 1 C-130.	Berger, 1984, p.182
454	5/23/72	Bien Hoa	Harass	Standoff attack (4 rds); ac losses 0/0.	Fox, 1979, p.201
455	5/25/72	Kontum	Deny defender use of airfield	NVA forces captured part of airfield, closing it.	Berger, 1984, p.182
456	6/4/72	Ubon, Thailand	Destroy aircraft	Sapper attack; ac losses 0/0.	USAF, 1973
457	6/10/72	Da Nang	Destroy aircraft	Standoff attack (8 rds); ac losses 0/2.	Fox, 1979, p.201
458	6/13/72	Da Nang	Destroy aircraft	Standoff attack (6 rds); ac losses 0/0.	Fox, 1979, p.201
459	6/17/72	Da Nang	Harass	Standoff attack (4 rds); ac losses 0/0.	Fox, 1979, p.201
460	6/22/72	Da Nang	Destroy aircraft	Standoff attack (6 rds); ac losses 0/0.	Fox, 1979, p.201
461	7/8/72	Da Nang	Destroy aircraft	Standoff attack (12 rds); ac losses 0/1.	Fox, 1979, p.201
462	7/13/72	Da Nang	Destroy aircraft	Standoff attack (16 rds); ac losses 0/0.	Fox, 1979, p.201
463	8/1/72	Bien Hoa	Destroy aircraft	Standoff attack (86 rds); ac losses 0/6.	Fox, 1979, p.202
464	8/3/72	Da Nang	Destroy aircraft	Standoff attack (45 rds); ac losses 0/4.	Fox, 1979, p.202
465	8/18/72	Da Nang	Destroy aircraft	Standoff attack (no. rds not available); damage in Event 464.	Fox, 1979, p.202

Table B.3—continued

Event	Date	Location	Objective	Description (aircraft [ac] losses destroyed/damaged)	Sources
466	8/18/72	Da Nang	Destroy aircraft	Standoff attack (35 rds); ac losses 2/10.	Fox, 1979, p.202
467	8/19/72	Da Nang	Harass	Standoff attack (2 rds); ac losses 0/0.	Fox, 1979, p.202
468	8/31/72	Bien Hoa	Destroy aircraft	Standoff attack (64 rds); ac losses 1/11.	Fox, 1979, p.202
469	9/10/72	Bien Hoa	Destroy aircraft	Standoff attack (1 rd); ac losses 3/95. Evidence suggests that one mortar round landed in Vietnamese Air Force munitions storage area, causing massive explosions and fires. Also possible that it was caused by sabotage or an accident in munitions handling.	Fox, 1979, p.202
470	9/10/72	Tan Son Nhut	Harass	Standoff attack (3 rds); ac losses 0/0.	Fox, 1979, p.202
471	9/23/72	Da Nang	Destroy aircraft	Standoff attack (27 rds); ac losses 0/3.	Fox, 1979, p.202
472	9/27/72	Da Nang	Destroy aircraft	Standoff attack (5 rds); ac losses 0/3.	Fox, 1979, p.202
473	10/22/72	Bien Hoa	Destroy aircraft	Standoff attack (56 rds); ac losses 0/7.	Fox, 1979, p.202
474	10/25/72	Da Nang	Destroy aircraft	Standoff attack (18 rds); ac losses 0/0.	Fox, 1979, p.202
475	10/28/72	Da Nang	Destroy aircraft	Standoff attack (27 rds); ac losses 0/8.	Fox, 1979, p.202
476	11/12/72	Bien Hoa	Destroy aircraft	Standoff attack (no. rds not available); damage in Event 475.	Fox, 1979, p.203
477	11/12/72	Bien Hoa	Destroy aircraft	Standoff attack (21 rds); ac losses 1/3.	Fox, 1979, p.203
478	11/19/72	Da Nang	Destroy aircraft	Standoff attack (9 rds); ac losses 0/2.	Fox, 1979, p.203
479	11/21/72	Da Nang	Harass	Standoff attack (4 rds); ac losses 0/0.	Fox, 1979, p.203

Table B.3—continued

Event	Date	Location	Objective	Description (aircraft [ac] losses destroyed/damaged)	Sources
480	12/1/72	Bien Hoa	Destroy aircraft	Standoff attack (28 rds); ac losses 0/14.	Fox, 1979, p.203
481	12/4/72	Bien Hoa	Destroy aircraft	Standoff attack (7 rds); ac losses 0/3.	Fox, 1979, p.203
482	12/6/72	Tan Son Nhut	Destroy aircraft	Standoff attack (28 rds); ac losses 0/2.	Fox, 1979, p.203
483	12/15/72	Bien Hoa	Destroy aircraft	Standoff attack (12 rds); ac losses 0/0.	Fox, 1979, p.203
484	12/16/72	Bien Hoa	Destroy aircraft	Standoff attack (6 rds); ac losses 0/0.	Fox, 1979, p.203
485	12/16/72	Bien Hoa	Harass	Standoff attack (3 rds); ac losses 0/0.	Fox, 1979, p.203
486	12/26/72	Da Nang	Destroy aircraft	Standoff attack (32 rds); ac losses 0/8.	Fox, 1979, p.203
487	1/14/73	Da Nang	Destroy aircraft	Standoff attack (7 rds); ac losses 0/4.	Fox, 1979, p.203
488	1/17/73	Da Nang	Destroy aircraft	Standoff attack (21 rds); ac losses 0/3.	Fox, 1979, p.203
489	1/22/73	Bien Hoa	Destroy aircraft	Standoff attack (10 rds); ac losses 0/1.	Fox, 1979, p.203
490	1/26/73	Bien Hoa	Destroy aircraft	Standoff attack (26 rds); ac losses 1/1.	Fox, 1979, p.204
491	1/26/73	Da Nang	Destroy aircraft	Standoff attack (12 rds); ac losses 0/0.	Fox, 1979, p.204
492	1/27/73	Da Nang	Destroy aircraft	Standoff attack (25 rds); ac losses 1/18.	Fox, 1979, p.204
493	1/28/73	Tan Son Nhut	Destroy aircraft	Standoff attack (11 rds); ac losses 0/0.	Fox, 1979, p.204

Table B.4

Chronology of Ground Attacks on Air Bases: Other Modern Conflicts and Terrorism

Event	Date	Location	Objective	Description	Sources
1	12/24/79	Kabul airport, various bases, Afghanistan	Seize airport as airhead	Soviet 105th Guards Airborne Division, elements of the 103rd Guards Airborne Division, and a Spetsnaz unit landed at Kabul airport and Bagram Air Base.	Middleton, 1979, pp.A1, A13; Collins, 1986, p.78
2	1/12/81	Muñiz Airport, Puerto Rico	Destroy aircraft	*Macheteros* terrorists raided Muñiz Air National Guard Base adjacent to International Airport at Verde. Using time-delayed bombs, they destroyed 8 A-7Ds and 1 F-104. Also damaged 2 A-7Ds & nearby equipment. Three unexploded bombs found on other aircraft. Total damage estimated at $45 million. Guard force consisted of one man at gate, another patrolling the perimeter. Attackers escaped undetected. (F-104 was non-operational & on static display.)	Thomas, 1981, pp. A1, A12
3	1/27/82	Illopango AFB, El Salvador	Destroy aircraft	Faribundo Marti National Liberation Front (FMLN) guerrillas attacked Illopango AFB with rockets and sappers in 2-hour-long predawn operation. 5 helicopters, 5 fighter aircraft, and 5 C-47 transports were destroyed. An additional 7 aircraft were damaged.	*Facts on File*, 1982

Table B.4—continued

Event	Date	Location	Objective	Description	Sources
4	5/14/82	Pebble Island, Falklands	Destroy aircraft	SAS detachment used satchel charges to destroy or damage 11 aircraft. Raid knocked out 30 percent of Argentine light aircraft in Falklands; airstrip remained out of action for rest of war.	Ethell, 1983, pp.65–66; Hastings, 1983, pp.186–187; Strawson, 1984, pp.231–232
5	10/25/83, 0500 local	Pearls, Grenada	Seize airport as airhead	U.S. Marines conducted airmobile assault and captured Pearl airport.	Adkin, 1989, p. 236
6	10/25/83, 0534–0707 local	Salinas, Grenada	Seize airport as airhead	U.S. 75th Rangers conducted airborne assault and captured Salinas airport. 82nd Airborne Division was airlanded, beginning 9 hours later.	Adkin, 1989, pp.200, 214, 217
7	5/27/86	Shindad Air Base, Afghanistan	Destroy aircraft	Afghan guerrillas claimed to have shot down a large Soviet transport aircraft on approach to Shindad with a SAM-7 man-portable missile.	Renfrew, 1986
8	5/30/86	Shindad Air Base, Afghanistan	Destroy aircraft	Afghan guerrillas struck the large Soviet air base in Western Afghanistan with 60 107-mm rockets in an attack that lasted 25 minutes. The attack destroyed 2 jet fighters and 6 helicopters caught in the open. A large fuel tank was also hit; it burned for 2 days.	Renfrew, 1986

Table B.4—continued

Event	Date	Location	Objective	Description	Sources
9	9/3/89	Monteria, Columbia	Destroy aircraft	U.S. State Dept. aircraft involved in anti-drug work was firebombed.	RAND Terrorism Database
10	12/20/89 (0020 local)	Albrook AFS, Panama	Harass	Unknown gunmen fired light weapons at USAF hangar. No damage or injuries. Gunfire came from tall grass near airfield perimeter fence.	McConnell, 1991, p.112
11	12/20/89 (0051 local)	Paitilla, Panama	Destroy aircraft	54-man SEAL (sea/air/land Navy special forces) detachment inserted from sea was detected and engaged by Panamanian Defense Force (PDF) in buildings. SEALs suffered heavy losses, but defeated PDF and destroyed Noriega's Learjet.	McConnell, 1991, pp. 47–72, 219
12	12/20/89 (0055 local)	Tocumen AFB/Torrijos Intl, Panama	Seize airbase and airport as airheads	1st Battalion & 1 co. 3rd Bat., 75th Ranger Regiment conducted airborne assault, capturing airfields and neutralizing PDF. 1st Brigade, 82nd Airborne Division, followed 1–4 hours later, also by airborne assault.	McConnell, 1991, pp.99, 191
13	12/20/89 (0057 local)	Rio Hato, Panama	Seize airfield as airhead	2nd Battalion, 75th Rangers conducted airborne assault, capturing airfield.	McConnell, 1991, p.73
14	11/20/90	San Salvador, El Salvador	Destroy aircraft	FMLN guerrillas attacked the major air base outside of San Salvador, damaging 1 aircraft and a barracks building.	Associated Press, 1990

Table B.4—continued

Event	Date	Location	Objective	Description	Sources
15	2/27/91	Jalibah AB, Iraq	Destroy ground forces at airfield	Three battalions from U.S. 24th Infantry Division attacked Jalibah Air Base. Tanks destroyed hangars, fuel tanks, helicopters, and 20 jet fighters. An enemy tank battalion was also destroyed.	Atkinson, 1993, p.455
16	2/27/91	Talil AB, Iraq	Destroy ground forces at airfield	A tank battalion from the 24th Infantry Division attacked Talil, destroying 6 MiGs, 2 helicopters, and a cargo plane.	Atkinson, 1993, p.455
17	3/17/91	Muñiz Airport, Puerto Rico	Destroy aircraft	Terrorists set fire to 1 A-7, causing $100,000 damage. Attackers escaped.	*Los Angeles Times*, 1991, p. A15
18	3/28/91	Khalid Air Base, Iraq	Destroy aircraft	Kurdish insurgents attacked Iraq Air Base at Khalid, southwest of Kirkuk, destroying 3 Su-22 jet fighters in hardened shelters and 4 Mi-8 helicopters.	UPI report, 1991
19	11/5/92	Isabela Prov., Philippines	Destroy aircraft	100 guerrillas attacked a Philippine air force base in the northern province of Isabela, destroying 2 OV-10 Bronco aircraft and damaging a Sikorksy S76 helicopter. This was the first time the NPA succeeded in attempts to destroy government aircraft.	Agence France Presse, 1992

BIBLIOGRAPHY

WORLD WAR II

Barnett, Correlli, *Engage the Enemy More Closely: The Royal Navy in the Second World War,* New York: W. W. Norton and Co., 1991.

Beevor, Antony, *Crete: The Battle and the Resistance,* Boulder, Colo.: Westview Press, 1994.

Bekker, Cajus, *The Luftwaffe War Diaries,* Garden City, N.Y.: Doubleday, 1968.

Blau, George E., *The German Campaigns in the Balkans (Spring 1941),* Washington, D.C.: U.S. Army Center for Military History, 1953, pp. 119–147.

Chennault, Claire Lee, *Way of a Fighter,* New York: G. P. Putnam's Sons, 1949.

Churchill, Winston, *The Second World War, Vol. III: The Grand Alliance,* Boston, Mass.: Houghton Mifflin, 1985 ed.

———, *The Second World War, Vol. V: Closing the Ring,* Boston, Mass.: Houghton Mifflin, 1951.

Clark, Alan, *The Fall of Crete,* New York: William Morrow and Co., 1962.

Cluxton, Donald E., Jr., "Concepts of Airborne Warfare in WWII," Master's Thesis, Duke University, Durham, N.C., 1967.

Cooper, Johnny, *One of the Originals: The Story of a Founder Member of the SAS,* London: Pan Books, 1991.

Craven, Wesley F., and James L. Cate, eds., *The Army Air Forces in World War II, Volume II,* Chicago, Ill.: University of Chicago Press, 1949.

———, *The Army Air Forces in World War II, Volume IV,* Chicago, Ill.: University of Chicago Press, 1950.

———, *The Army Air Forces in World War II, Volume V,* Chicago, Ill.: University of Chicago Press, 1953.

Crichton-Stuart, Michael, *G Patrol,* London: William Kimber, 1958.

Davin, D. M., *Crete: Official History of New Zealand in the Second World War 1939–45,* Wellington, New Zealand: War History Branch, Department of Internal Affairs, 1953.

Fuller, J. F. C., *The Second World War, 1939–45,* New York: Duell, Sloan and Pearce, 1948.

Gordon, John W., *The Other Desert War: British Special Forces in North Africa, 1940–1943,* New York: Greenwood Press, 1987.

Greiss, Thomas E., ed., *The Second World War: Asia and the Pacific,* The West Point Military History Series, Wayne, N.J.: Avery Publishing Group, Inc., 1984a.

———, *The Second World War: Europe and the Mediterranean,* The West Point Military History Series, Wayne, N.J.: Avery Publishing Group, Inc., 1984b.

———, *Atlas for the Second World War: Asia and the Pacific,* The West Point Military History Series, Wayne, N.J.: Avery Publishing Group, Inc., 1985a.

———, *Atlas for the Second World War: Europe and the Mediterranean,* The West Point Military History Series, Wayne, N.J.: Avery Publishing Group, Inc., 1985b.

Grey, C. G., *The Luftwaffe,* London: Faber and Faber, 1944.

Hart, B. H. Liddell, *History of the Second World War,* New York: G. P. Putnam's Sons, 1970.

———, ed., *The Rommel Papers,* London: Collins, 1953.

Hoe, Alan, *David Stirling: The Authorized Biography of the Creator of the SAS,* London: Warner Books, 1992.

James, Malcolm, *Born of the Desert,* London: Greenhill Books, 1991.

Kay, R. L., *Long Range Desert Group in Libya: 1940–41,* Wellington, New Zealand: War History Branch, Department of Internal Affairs, 1949.

———, *Long Range Desert Group in the Mediterranean,* Wellington, New Zealand: War History Branch, Department of Internal Affairs, 1950.

Keegan, John, *The Second World War,* New York: Viking, 1989.

Kreis, John F., *Air Warfare and Air Base Air Defense,* Washington, D.C.: U.S. Air Force Office of History, 1988.

Ladd, James D., *SBS: The Invisible Raiders, The History of the Special Boat Squadron from World War II to the Present,* Annapolis, Md.: Naval Institute Press, 1983.

Lewin, Ronald, *Ultra Goes to War,* New York: McGraw-Hill, 1978.

Lodwick, John, *Raiders from the Sea: The Story of the Special Boat Service in WWII,* Annapolis, Md., 1990.

Lucas, James, *Kommando: German Special Forces of World War Two,* New York: St. Martin's Press, 1985.

Messenger, Charles, *The Commandos: 1940–1946,* London: William Kimber, 1985.

Murray, Williamson, *Strategy for Defeat: The Luftwaffe 1933–1945,* Maxwell AFB, Ala.: Air University Press, 1983.

Oliver, Kinsley M., *A Short History of the RAF Regiment,* Church Hill, Ramsgate, England: Thanet Printing Works, 1970.

Owen, David Lloyd, *Providence Their Guide: A Personal Account of the Long Range Desert Group, 1940–45*, London: Harrap, 1980.

Peniakoff, Vladimir, *Popski's Private Army*, New York: Thomas Y. Crowell Company, 1950.

Richards, Denis, *The Fight at Odds, Volume I, Royal Air Force 1939–1945*, London: Her Majesty's Stationery Office, 1953.

Romanus, Charles F., and Riley Sunderland, *United States Army in World War II, China-Burma-India Theater: Stilwell's Command Problems*, Washington, D.C.: Department of the Army, 1956.

Saunders, Hilary St. George, *Royal Air Force 1939–1945, Volume III, The Fight Is Won*, London: Her Majesty's Stationery Office, 1954.

Shaw, W. B. Kennedy, *Long Range Desert Group: The Story of Its Work in Libya, 1940–43*, London: Collins, 1945 (1989 ed.).

Steward, Ian, *The Struggle for Crete*, London: Oxford University Press, 1966.

Strawson, John, *The Battle for North Africa*, New York: Charles Scribner's Sons, 1969.

———, *A History of the S.A.S. Regiment*, London: Secker and Warburg, 1984.

Tucker, Nick, "In Adversity: Exploits of Gallantry and Awards in the RAF Regiment and Its Associated Forces," unpublished manuscript.

Van Creveld, Martin, *Supplying War: Logistics from Wallenstein to Patton*, Cambridge, England: Cambridge University Press, 1979.

Verney, John, *Going to the Wars*, London: Collins and the Book Society, 1955.

Warner, Philip, *The Special Air Service*, London: William Kimber, 1971.

Winterbotham, F. W., *The Ultra Secret*, New York: Harper and Row, 1974.

KOREA

Futrell, Robert F., *The United States Air Force in Korea, 1950–1953,* Washington, D.C.: U.S. Air Force Office of History, 1983.

Schuetta, Lawrence V., *Guerrilla Warfare and Airpower in Korea, 1950–53,* Maxwell AFB, Ala.: Aerospace Studies Institute, 1964.

VIETNAM

Berger, Carl, ed., *The United States Air Force in Southeast Asia, 1961–1973,* Washington, D.C.: U.S. Air Force Office of History, 1984.

Fox, Roger P., *Air Base Defense in the Republic of Vietnam: 1961–1973,* Washington, D.C.: U.S. Air Force Office of History, 1979.

Francillon, Rene, *Vietnam: The War in the Air,* New York: Arch Cape Press, 1987.

Palmer, Bruce, *The 25 Year War: America's Military Role in Vietnam,* Lexington, Ky.: University Press of Kentucky, 1984.

United States Air Force (USAF), *Attack Against Tan Son Nhut: Project CHECO Southeast Asia Report,* Hickam AFB, Hawaii: Headquarters, Pacific Air Forces, July 8, 1966.

———, *Attack on Udorn (26 July 1968): Project CHECO Southeast Asia Report,* Hickam AFB, Hawaii: Headquarters, Pacific Air Forces, December 17, 1968.

———, *7th Air Force Local Base Defense Operations (July 1965–December 1968): Project CHECO Southeast Asia Report,* Hickam AFB, Hawaii: Headquarters, Pacific Air Forces, July 1, 1969a.

———, *Defense of Da Nang: Project CHECO Southeast Asia Report,* Hickam AFB, Hawaii: Headquarters, Pacific Air Forces, August 31, 1969b.

———, *Local Base Defense in RVN (January 1969–June 1971): Project CHECO Southeast Asia Report,* Hickam AFB, Hawaii: Headquarters, Pacific Air Forces, September 14, 1971a.

———, *Attack on Cam Ranh (25 August 1971): Project CHECO Southeast Asia Report,* Hickam AFB, Hawaii: Headquarters, Pacific Air Forces, December 15, 1971b.

———, *Base Defense in Thailand: Project CHECO Southeast Asia Report,* Hickam AFB, Hawaii: Headquarters, Pacific Air Forces, February 18, 1973. (Declassified by USAF Office of History, August 16, 1994.)

———, *USAF Management Summary: Southeast Asia Review,* Hickam AFB, Hawaii: Headquarters, Pacific Air Forces, February 28, 1974. (Declassified by USAF Office of History, 1994.)

OTHER MODERN CONFLICTS

Adkin, Mark, *Urgent Fury: The Battle for Grenada,* Lexington, Mass.: Lexington Books, 1989.

Agence France Presse, "Communist Guerrillas Destroy Two Air Force Planes," dateline: Manila, Philippines, November 6, 1992.

Associated Press, "Salvadoran Rebels Hit Military Posts," *Chicago Tribune,* November 21, 1990, p. 3.

Atkinson, Rick, *Crusade: The Untold Story of the Persian Gulf War,* New York: Houghton Mifflin, 1993.

Burden, Rodney A., et al., *Falklands: The Air War,* Dorset, England: Arms and Armour Press, 1986.

Collins, Joseph J., *The Soviet Invasion of Afghanistan,* Lexington, Mass.: Lexington Books, 1986.

Donnelly, Thomas, Margaret Roth, and Caleb Baker, *Operation Just Cause: The Storming of Panama,* Lexington, Mass.: Lexington Books, 1991.

Ethell, Jeffrey, and Alfred Price, *Air War South Atlantic,* New York: Macmillan Publishing Co., 1983.

Flanagan, Edward M., *Battle for Panama: Inside Operation Just Cause,* New York: Brassey's, 1993.

"Guerrilla Attacks Intensify," *Facts on File,* February 5, 1982.

Hastings, Max, and Simon Jenkins, *Battle for the Falklands,* New York: W. W. Norton and Co., 1983.

Hoffman, Bruce, *Recent Trends and Future Prospects of Terrorism in the United States,* Santa Monica, Calif.: RAND, R-3618, 1988.

"Intruders Damage Plane at U.S. Base in Puerto Rico," *Los Angeles Times,* March 18, 1991, p. A15.

McConnell, Malcolm, *Just Cause: The Real Story of America's High-Tech Invasion of Panama,* New York: St. Martin's Press, 1991.

Middleton, Drew, "Soviet Display of Flexibility: Afghan Airlift Is Lesson in Moving Troops Fast," *The New York Times,* December 28, 1979, pp. A1, A13.

RAND, Terrorism Database, Santa Monica, Calif., n.d.

Renfrew, Barry, "Guerrillas Report Attack on Major Soviet Air Base," Associated Press Report, dateline: Islamabad, Pakistan, June 8, 1986.

Richissin, Todd, and Ben Stocking, "Peace Activists Held for Attack on Military Jet," *Raleigh News and Observer,* December 8, 1993, p. 3.

Thomas, Jo, "Puerto Rico Group Says It Struck Jets," *The New York Times,* January 13, 1981, pp. A1, A12.

United Press International, "Kurdish Guerrillas Attack Air Base, Destroy Aircraft," dateline: Athens, Greece, March 28, 1991.